In the Shadows Between the Wars

Becoming a Special Forces Operator

Christopher Brewer

In the Shadows Between the Wars:
Becoming a Special Forces Operator

Copyright © Christopher Brewer (2025)

Published by:

CLEARED
For Open Publication
Aug 22, 2024
Department of Defense
OFFICE OF PREPUBLICATION AND SECURITY REVIEW

The views expressed in this publication are those of the author and do not necessarily reflect the official policy or position of the Department of Defense or the U.S. Government.

The public release clearance of this publication by the Department of Defense does not imply the Department of Defense endorsement or factual accuracy of the material.

This book is dedicated to my son, Eric, who had a tough life growing up with a Dad gone much of the time.

And to the good people who helped us through those difficult years, my brother Todd and my best friends Mike and Joanne.

Table of Contents

Introduction

This story takes a bit of explaining. It's not your typical Special Forces "now it can be told" book. It all got started when I saw the books on the Studies and Observation Group and Project Phoenix from Vietnam coming out. Those were Top Secret, never to be told, take it to your grave stories that now have their own Face Book pages.

I know some of those guys, and I thought, what the hell, I've got a few of those "Don't you ever talk about this" stories of my own. It appears that after twenty years or so, DOD is willing to let the cat of the bag but you have to ask first.

So I asked.

DOD wrestled with it for more than six months, sent it back and forth to different agencies and after much soul searching sent it back with a surprisingly small number of redactions. They insisted I change an acronym or two and not mention one or two things but for the most part they let it all go.

But this is not a "tell all" book about secret missions. There are some stories here that were Top Secret and some that were Special Category that are worth reading and haven't been told before. But there are people and units that are not mentioned so the essence of the story can be presented. If you get the sense there was more to the story, you are probably correct. You figure it out.

It's not a comprehensive history of Special Forces during that time, although it covers some historical events that had a major impact on SF. What I saw was not the whole picture.

What it is is a first-person perspective of what one person went through on the way to SF and what I experienced while there. It starts with a short story about trying to be a civilian and deciding I would rather be a soldier. No one makes it into SF unless they really want to be there.

The first couple of chapters after that cover what motivated me to try SF, the challenges encountered getting there, and some events that prepared me for what would follow. I picked ones that I hope will be entertaining and capture some of the history of other units that touch on SF operations and how we operated in the 70's and 80's. The 509th Airborne Combat Team in Vicenza, Italy, and Platoon Confidence Training at Bad Tolz under the 1/10th Special Forces Group, to be specific.

From there, you will get a sense of what happened to SF when it became a branch, when Selection and Assessment was born, and what it's like to go through the Qualification Course.

And then we get into the secret stuff. This covers ODA 745 in Colombia and Panama after the drug cartels started an open war with the Colombian government, Operation Promote Liberty after Operation Just Cause, and 1/1 SFG in Okinawa. I had fourteen years in the Army before I started the Q course, but three years of that were in Special Forces units before the Q Course and another nine years after I was qualified. That gave me a little different perspective.

These are stories of what I experienced, but it's not all about me. It's about the soldiers in Special Forces, what they go through to get there, and a bit of what they experience on the teams. These are their stories, too. I have not used people's names out of respect for their privacy, but they know who they are. This is for them so they can show their kids some of the things they did we were told not to talk about.

It's also for the soldiers in Special Forces now. I've talked to guys currently serving in the teams that I had the privilege of being a part of who have no idea what their team history includes. The stories get lost with time, and some of the lessons learned as well. Hopefully, they will get something from these pages that they can use.

I was a "tweener," a soldier who served after Vietnam and retired before 911. I didn't participate in any of the major wars of the time. Although I was on active duty during Operation Urgent Fury, Just Cause, Desert Shield, and Desert Storm, I didn't have an opportunity to deploy to any of those conflicts—at least not when they were handing out merit badges. We got shot at occasionally, but nothing like what the guys experienced every day in the conflicts after 911.

Most of what we went through was a constant tension of knowing you were standing next to someone who wanted you dead, was planning how they could get away with it, and you had to prevent that from happening while trying to get them to work with you. Believe me, it's a lot simpler when you can just shoot them.

They used to call the areas where conflicts occurred "combat zones." Basically, it means a place where large groups of people organized into units are fighting and killing one another for one reason or another. Special Forces is always present when big Army goes to war in named Operations.

But SF also goes to other nations' wars and dangerous areas that are not full-blown combat but dangerous nonetheless. Often, it is not our fight, and the U.S. is not officially at war with either side, but it is in our interest to influence the outcome. These missions are often classified to keep them low profile and out of the newspapers.

It isn't like the CIA, although we often run into them doing their thing while we do ours. The people we work with and their enemies know who we are and that we are there in most cases. SF is there for the long fight, to help the host nation or the people fighting for their freedom.

"By, with, and through" host nation forces is the mantra of Special Forces. U.S. Special Forces' mission is to enable, train, and support the host nation's forces to fight their battles instead of us fighting for them. That can be difficult, and it's always dangerous.

It means living with the host nation forces instead of huddled in a Forward Operating Base defended by U.S. troops with a PX, Pizza Hut, and Dairy Queen. You eat what the local forces eat, live where they live, and face the same dangers they face. Sometimes, the enemy is hidden in their forces. We needed to avoid killing anyone ourselves since that could turn into a news event. On occasion, it happened anyway. "You were not there, and it didn't happen" were things we heard more than once.

It's worth mentioning here the much-overused term "operator." It seems everyone is an operator these days. The term originated in World War II and described the OSS men and women operating behind enemy lines. It means the person is operating alone or as part of a small group in an area where the usual chain of command and support mechanisms found in large military organizations at war simply do not exist.

The operator is on their own. They go out with a mission and instructions, but they will encounter unexpected situations where "things go sideways," the original plan no longer works, or the objective is no longer relevant. The operator must decide what course of action to take and move forward without approval from higher headquarters. No one is coming to help them, and it's usually not possible to consult with higher for guidance either because of lack of time or inability to communicate.

In the 1980s and 90s, the ubiquitous instant communication systems of the Internet, cell phones, and satellite communications did not exist or were in the early stages of development and fielding. We were instructed to try to "make coms" once a month so our commanders would know we were still alive and we could exchange notes on what was going on. Sometimes, we had access to landline telephones; often, we did not.

This sets Special Operations apart from conventional units and makes conventional commanders nervous. Often, it's up to the operator to make decisions that will strategically impact U.S. policy. If it's a good decision, usually someone else gets credit; if not, it's written off as the actions of a "loose cannon."

During the period between the Vietnam War and the wars in the desert, Special Forces were deployed in over one hundred and thirty-two nations across the globe, often engaged in active combat, sometimes suffering fatal casualties, but sworn to secrecy. Our wars were in the shadows, and we were told they would always stay there.

The young soldiers entering service now may worry that they will never get to prove themselves in combat against an adversary. We are now, perhaps, done with the large-scale deployments of conventional military units and moving into what hopefully will become another "tweener" period of relative peace. We'll see how long that lasts.

But the war in the shadows never ends.

CHAPTER ONE

Transition

You have twenty seconds to live.

T hat's what your reality becomes when you step out of an aircraft three thousand feet above the ground.

Unless you do something about it and do everything right.

As you fall, it takes twelve seconds to build up speed. You will fall 1,500 feet in the first twelve seconds.

After twelve seconds, you are at terminal velocity, the highest average speed a human body will fall through the air in a normal, stable body position. You are now falling at one hundred and seventy-six feet per second, and you will hit the ground at one hundred and twenty miles per hour in eight and half seconds.

These are important things to know before you decide to take that leap. And there are preparations you need to make before you do so.

Like wearing a parachute—ideally, a main and a reserve—so you have two chances to survive the fall. Believe it or not, there are recorded instances of people getting in a rush and forgetting to do that.

And finally, you need to remember that the situation will only become more dangerous and difficult as you fall. It's a major transition from solid earth to thin air, so you better be ready for what's coming.

That happened to me more than once throughout the following story, literally and metaphorically.

In 1977, I left the Ranger Battalion and the Army. I wasn't ready, and I wasn't prepared. The Batt was going through a slump; standards were dropping, and the old guard was leaving for other units in the stated mission of the battalion to go forth and raise the Army's standards as a whole. The Ranger battalion slump was temporary, and it would come roaring back better than before, but I had become disillusioned with Army life. I grew up an Army brat, went to a military college in North Georgia, and then to the Ranger battalion, so most of my life was Army up to that point, and I figured it was time to try out civilian life.

It wouldn't last more than a year.

Before I became a Ranger, I had been an Army Medic, so I tried being an emergency room orderly. But an orderly isn't allowed to do much more than clean and bring supplies and instruments to the doctors and nurses, so there wasn't much of a challenge. And seeing sick and injured people flooding in every day, all day, and not being able to help them takes a toll on your spirit.

I moved to California to visit my Dad and see the West Coast. I did several odd jobs and worked for a guy who owned a burglar alarm company in Sausalito. He wanted to branch out into bodyguard service, so I became his first gun for hire. Except I didn't have a California gun license or a bodyguard license, for that matter. It didn't matter to Tommy (names changed to protect the guilty); he didn't have any licenses either.

Tommy had a New York accent, a distinctly Italian look and appetites but an old English last name. He had been in California for a couple of years but denied ever being in New York and didn't want to talk about anything before living in California. You can guess what that was about, and you would probably be right, but it was OK with me; it was interesting work.

I worked in San Francisco, Sausalito, and Oakland, primarily unarmed, escorting businessmen or guarding their homes or offices in uniform and plain clothes. Met a lot of interesting people. Until the night Tommy told me to meet him at an address in Sausalito. I was supposed to guard an accountant who had to testify in court the next day and had been threatened if he testified. It was pretty simple, just keep the client from being harassed at home or beat up the next day on the way to court.

At around 10:00 that night, I met Tommy at the prescribed address, rang the bell, and waited for someone to answer. After a minute, I heard Tommy whisper from behind the locked door, "Who's there?"

"It's me, Chris."

"Step back into the light so I can see your face! Are you alone?!?!"

"Yeah, Tommy, open up, what's the deal?"

The door opened, and Tommy pulled me inside quickly. He had a pump shotgun in his hand, and he handed me a .25 automatic Saturday night special pistol.

I stood there momentarily looking at the pop gun Tommy had just handed me and figured there was more to this gig than he had told me.

There was.

Tommy said, "I didn't know this myself until I got here, and the client started talking. I didn't believe it at first, but I made some phone calls, and as far as I could tell, he was telling the truth."

"This guy is an accountant for some mob guys in Miami, and he was holding a couple of hundred thousand dollars for them to deliver to one of their guys here."

"But the cops found out the accountant had the money, impounded it, and made a deal to give him immunity if he testified in court about who it came from and who he was supposed to give it to."

"The mob guys in Miami don't know that the accountant has agreed to testify, but when he didn't deliver the money on time, their guy in San Francisco called them. Their guy here told them he didn't deliver the money, and the accountant was afraid to tell them the cops had it, so he said he did give it to their guy. The mob told him they were sending an enforcer to find out who was lying to them."

"So, a mob guy here in San Francisco is looking for the accountant, and an enforcer from Miami is looking for the accountant and the mob guy here."

You've got to be kidding me. "Tommy, I'll do this tonight, but I'm done once we get this guy to the courthouse tomorrow. We are not equipped to deal with this, and you damn sure aren't paying me enough."

Lord, have mercy. If I have to throw down with the mob, I might as well be back in the Army. At least there, they give me better guns.

The night passed without incident, and we handed the accountant over to the police.

But I was headed back to the big green machine. I had jumped before I considered all the angles, and the ground was coming up fast; it was time to pull my reserve and abort. The recruiter offered me a slot in the battalion, but I was still not happy with what I had seen there when I left, so I opted for something different.

Since I seemed to be dealing with Italians, I might as well go to the source. I re-enlisted for the 1st/509th Airborne Battalion Combat Team in Vicenza, Italy.

CHAPTER TWO

1/509th Airborne Battalion Combat Team

The 509th unit has a proud history, and the guys who served in it are proud of it. The unit traces its lineage to the 504th Parachute Infantry Regiment, the first airborne unit to conduct a combat jump in Africa. The unit took heavy casualties during the war and was often reorganized, redesignated as the 503rd and finally the 509th. They conducted a total of five combat jumps during World War II. The unit was deactivated after its last mission with over 90 percent casualties. Its surviving members were transferred to the 82nd and 17th Airborne Divisions. The unit was reactivated in 1963 in Mainz, Germany, as part of the 8th Infantry Division. In 1973, the unit was moved to Vicenza, Italy.

As the only U.S. Army airborne infantry unit in Europe, the guys worked hard to live up to the legend. They worked hard, drank hard, and sometimes fought for the sheer hell of it. It was a good unit, and its soldiers were good at what they did, keeping the spirit of the World War II paratroopers alive.

But there were challenges. To fully appreciate the situation, you need to understand what the Army was like in the '70s.

In 1978, Vietnam had been over for about four years. All the factors of poor discipline, drug use, and racial conflict that led to the formation of the Ranger Battalion were still there in the regular Army. The 509th was better than most U.S. units in Germany in this regard, but they had some of their own issues.

Drug use was common, with hashish a drug of choice due to its availability. Marijuana wasn't grown in Europe, but it was in Africa and the Middle East, where it was processed into hashish and smuggled in through a variety of routes, many of them through Italy.

Hashish was illegal and punished when troopers were caught, but there was no urinalysis back then. The Army followed the "zero tolerance" policy. Get caught with any drugs, including marijuana or hashish, and you will be thrown out of the Army on a less-than-honorable discharge. But that didn't happen fast; it usually took them six months or better to process the charges and the discharge. That left a very unhappy paratrooper who really didn't have anything left to lose and nothing to gain by following the rules. In short, it created more problems than it fixed.

On top of that, alcohol was not just legal; consumption was expected. There were soda machines and beer machines in the barracks. You were allowed to have two beers with lunch during duty. And if you were an officer or senior Non-Commissioned Officer (NCO), you were expected to show up after work at the Officers or NCO Club to tip a few with your fellow leaders. Which often led to more than a few.

Between drugs and alcohol and a lot of spirited paratroopers who wouldn't walk away from a fight, it could be a challenge for the unit leadership.

Minor issues like being late for formation, insubordination, or refusing to obey orders were usually handled by the unit's NCOs. That includes three-striped Sergeants and Staff Sergeants who work at the squad level, Sergeant First Class at the platoon, and the First Sergeant of the company. Like I said, the majority of the soldiers in the unit were good men who might get in a bit of trouble, but nothing we couldn't handle and correct at the NCO level.

And being in Italy presented its own challenges. As nice as it was to be in Bella Italia, being an American soldier there could be tough at times. We had two local places where we could go to the field to train: a dry riverbed and an old garbage dump. You couldn't ruck march off post since the Italian government didn't want U.S. soldiers walking through the countryside and along their roads carrying weapons. The military base was small, with nothing more than barracks, a parade field, and a football field.

Aldo Moro, the Prime Minister of Italy, had recently been kidnapped and murdered by the Red Brigade, and U.S. soldiers walking on the side of the road would be too easy of a target. Two years after I got there, U.S. Major General Dozier was kidnapped from his quarters by the Red Brigade. So training was whatever you could do on the base under normal circumstances, and actual training required an off-post deployment that usually lasted several weeks. This left hard-charging soldiers with a lot of time on their hands between deployments and opportunities to get into trouble.

We could only do local training on the football field or the company area around the barracks. At least once a year, we would deploy to Wildflicken or Hohenfels, where we did most of our weapons qualification and tactical training. That was some of the best training, and often the entire battalion deployed.

Charlie Company deployed to Spain when I was there and set a precedent for future deployments for joint training with the Spanish airborne. It was excellent training in the Spanish desert that involved long foot marches and environmental training that would be useful for a mission in Africa.

We also deployed to Bad Tolz, Germany, for Platoon Confidence Training (PCT). We sent one platoon every cycle that had an opening. I was a member of the 20th Special Forces Group as a junior medic in the National Guard when I went to college, but that was as a "Candy Bar" green beret, someone not SF qualified but assigned to the unit and not authorized to wear a full flash on the beret. Candy Bars wore a cut-down version of the flash under the SF crest and performed limited duties they were qualified to perform. I worked as a basic skill medic in the battalion Head Quarters and didn't really see what SF did in the field.

PCT was my first exposure to some of what Special Forces did in the field. Certainly not everything, but a lot more than I saw in the 20th SFG. It made an impression. Two Special Forces Operational Detachment A's (ODAs) started PCT to practice one of the core Special Forces missions, organizing and training indigenous resistance forces.

Platoons from the U.S. Army units in Europe would travel to Bad Tolz and stay at Camp Warden for two weeks of training. The ODAs would conduct physical training and ruck marches, teach raid and ambush classes, and train the troops in mountaineering techniques. Then they would take the platoons to the field to conduct patrols in the German Alps around Flint Kaserne to practice what they had learned.

The interesting thing about Camp Warden and the teams that ran the training was that the Army did not officially recognize or fund it. The ODAs came up with the idea for PCT to practice training resistance forces as part of Unconventional Warfare.

They had no ammunition budget, barracks, or anything to support the effort. But being Special Forces they got creative, sold the idea to their chain of command, and through them to conventional U.S. Army units in Europe. The units would deploy to Bad Tolz with enough ammunition, rations, and other supplies to sustain themselves for the two weeks they would be there.

The SF teams convinced engineer units to come for training and bring building materials. As part of their training, the engineers built temporary barracks out of plywood and tar paper, graded roads, and built berms for live fire ranges.

Over time, the ODAs collected old tents, ammunition, explosives, and everything they needed to conduct training from leftovers collected throughout the European command. The U.S. Forces Europe commander heard about it and liked it so much that he directed the SF battalion to establish this as a full-time school and run classes all year round.

Now, the ODAs who started this knew it was great training for the platoons and the ODAs, but the problem was that this wasn't the only thing SF did. There were five doctrinal missions Special Forces teams are supposed to train to conduct: Unconventional Warfare, Direct Action, Foreign Internal Defense, Special Reconnaissance, and Counter Terrorism.

Each mission requires specialized training, and every SF team must set time aside to train for those missions. PCT took too much training time to support and still stay current in the other Special Forces missions.

So, the ODAs had another bright idea. SF has always had difficulty finding enough quality personnel who can make it through SF training to man the teams. Every SF battalion had a shortage of personnel, and 1/10th was no exception. Every ODA was usually short of two or three personnel. In addition, SF training can be high-risk. People get hurt, and sometimes, healing takes a long time.

The battalions can, and often do, designate "ghost teams" that are on the books, but no one is assigned. That allows them to fill twelve-man teams instead of having teams that are always short of personnel. When someone gets hurt or cannot deploy for training for an extended period, they are assigned to the ghost team and perform different functions.

Each battalion had three line companies, with six ODAs in each company. At the time, 1/10th had two ghost teams to allow them to keep the other teams up to strength. The battalion arranged with Personnel Command back in the US to assign personnel to the battalion who were Ranger, but not SF qualified if the battalion commander approved them. Personnel in ODAs who were injured or otherwise needed time with their families away from deployments were also assigned there.

This opened the door for NCOs from the Ranger battalion to get a coveted assignment to Bad Tolz. And after going through a PCT cycle, I wanted to be one of them. I was deeply impressed by Special Forces and I wanted to be part of it. This started a chain of events that led to it happening for me, but there were a lot of challenges that lay ahead.

For the moment, I was still a young three-stripe buck sergeant assigned to the 1/509th and getting used to the conventional Army. Each company in the 509th had a couple of Rangers, and the chain of command was glad to have us. But the Army's policy of throwing out soldiers for smoking pot or hash was pulling a lot of otherwise good soldiers into a pit they couldn't climb out of. Once charged, discharge would happen, but later, not sooner. Guys waiting for that to happen could be tough to motivate when they feel they have nothing to lose.

Sometimes, it was a matter of who was the baddest guy in the valley, regardless of rank. We called it "wall-to-wall counseling." If you were in charge, and someone told you to fuck off when you gave them an order, you could refer them for charges, and they would be confined to the barracks and maybe lose a rank. But if they were already pending discharge for drugs or other offenses, and they were already busted down to private, that didn't mean much to them. And you would still be stuck with them for six months while they waited to be processed out of the Army. Or, as an NCO, you could take things into your own hands and go "wall to wall" with the offending soldier.

To an extent, we had that system in the Ranger battalion, too. The difference was every NCO in the battalion (except one) had combat experience in the Long-Range Reconnaissance patrols in Vietnam, the hatchet companies, the Studies and Observation Group (SOG), or Special Forces, and every one of them was terrifying to a young private. Very rarely did an NCO in the Ranger batt have a problem with a private ignoring orders. But when it did happen, it was settled quickly, and the private always lost.

So, on my first week in the 509th I was on the duty roster for Charge of Quarters on a Saturday night. Things were usually pretty calm in the barracks during the duty day, but the animals would come out at night on the weekends. Some of the troopers would head downtown to the bars, drink heavily, maybe have a fight or two, and come back to the barracks well-lubricated. They traveled in packs and hung together. And some of them were not in the mood to follow orders, especially if one or two in the group were already awaiting discharge from the Army.

I came in at the appointed time on Saturday morning to pull my duty and looked around for the CQ from Friday night. Usually, he would be at a desk by the front door to control who came and went. But tonight, there was no desk, and no CQ was to be found.

I walked around, called out his name, and heard a frantic whisper from behind me in the hall. "Here, in the First Sergeant's (1SG) office. Get in here, quick!" I looked behind me and saw a very frightened buck sergeant staring around the 1SG's door with two black eyes.

I walked into Top's office and asked him what the hell was going on. He replied, "We always move the CQ desk into Top's office on the weekends so we can lock the door. But I went out last night to get a soda from the machine in the day room, and the privates caught me. I'm lucky to be alive."

I should mention here that he wasn't entirely wrong. When the 509th was in Germany before they moved down to Italy, a Lieutenant was making the rounds in the barracks checking security when he chanced upon a group of privates smoking hashish in their room. He sternly informed them that they were "under arrest" and ordered them to hand over the drugs and follow him to the Staff Duty Office so he could turn them over to the Military Police.

The problem with that was the troopers were drunk, stoned, and really didn't want to hand over their hash and be arrested. They beat the LT, stuffed him onto a wall locker, and locked it.

Now, the LT was in a pickle, but he didn't let that stop him from exerting his authority. Loudly and repeatedly, he demanded they let him out and surrender. The troopers got tired of hearing this and pushed the wall locker, with the LT inside, out the window. Since they were on the third floor, this had a negative impact on the wall locker, the sidewalk, and the LT.

The LT died. The troopers involved all went to the Mannheim confinement facility for murder.

But that was an extreme case. Most encounters only involved a beatdown.

I had heard the story. There were two other Ranger battalion sergeants in the company, and they had given me the low down on discipline in the 509th. The troopers would respect anyone who deserved it and could stand up to the troublemakers on their own. It was usually enough if you looked, acted, and conducted yourself as a professional and made it clear you wouldn't back down.

But once in a while, it wasn't. A group led by a troublemaker who just doesn't care can be a problem, especially if they are all drunk and high. We also had a very small number of less-than-effective NCOs who really shouldn't have been NCOs, but they allowed their troopers to walk all over them. A paratrooper who doesn't respect his leader can and sometimes will resort to violence and intimidation if he thinks he can get away with it.

You have to prove you can't be messed with; in short, "earn your street creds." That doesn't necessarily mean physical violence, but you need to make it apparent you won't back down, and there will be immediate repercussions if your orders are not followed. The best way is to lead from the front, set a good example, be physically fit, and be tactically and technically competent.

Paratroopers want to be soldiers and be led by good leaders, and they will respond. But if their NCO is a "ROAD warrior" (Retired On Active Duty) doing as little as possible to get by, the soldiers under him will be disgusted with him, the Army, and life in general, and problems will arise.

I had already been challenged by a few troopers who chuckled at the new buck sergeant, but I didn't have trouble with the guys in my squad, so I let it slide. No one had defied me, and the troublemakers were not my responsibility, so I took note of their attitude and moved on. But now I was the CQ, and the entire company was mine tonight.

I relieved the CQ with the two black eyes and pulled the desk back into the hall by the door. There was a steel triangle with a steel rebar on a chain hanging by the door used as a fire alarm, so I took the rebar off the chain, hefted it in my hand, and decided it would do for tonight's festivities. For good measure, I got a broken pool cue out of the day room, put both items on the desk, and waited for the first group of guys to come back from downtown.

The first group that walked in happened to include one of the troublemakers who had commented earlier in the week about "showing the new buck sergeant what's what in the '09". The same group had roughed up the other buck sergeant Friday night. He laughed when he saw me, reached across the desk, and grabbed my throat with both hands.

Game on.

A quick sweep with my left arm knocked his hands away from my throat. The pool cue in my right hand let everyone know this wasn't going down the same way as last night. I let the grabber know he would immediately regret it if he raised a hand to me again. He and his merry band were confined to their room for the duration of the night, and if I so much as saw their nasty little faces, I would rearrange their features to make my point. I picked up the steel rebar with my left and came out from behind the desk, shouting at them to back off and move upstairs, or bodies were about to hit the floor.

This elicited the desired response. They were drunk and not used to that kind of pushback, so they backed off and scurried upstairs, shouting, "OK, Sarge, OK, Jesus, chill out. We were just playing!"

Now, these days, that might result in charges against me for verbal abuse and threats of violence. But I wasn't making a threat, and if anyone had pressed forward with an attack, I would have defended myself. I made it very plain that the first person who got hurt wouldn't be me.

When applying violence, it's important to do so carefully, unemotionally, and purposefully. When the troopers came in, I didn't get aggressive; I simply smiled at them and waited to see what they would do.

Had they gone upstairs, even with some snide comments like those they made in our earlier encounter, I would have let it slide. No point in trying to confront them at night when I was outnumbered, and they were all drunk. That would inevitably have led to a confrontation. There is a time and place.

But when the ringleader put his hands on my throat, the equation changed immediately. That's grounds for self-defense, but more importantly, it's a breakdown in discipline. It has to be challenged and corrected immediately, or you lose all credibility. That said, hitting him with the rebar or in the face or with my fist would have caused visible and potentially serious injury and called my judgment into question. It would also probably have led to a fight with the whole group as they defended their ringleader.

But if he had pressed his attack, a gentle love tap on the side of the knee with a wood pool cue would be immediately painful, incapacitating, and wouldn't leave a mark. It would have made my point, and the ringleader would likely have backed off. He would have definitely hit the floor.

A painful but minor injury like that would motivate the rest of the crew to help him up and get the hell out of there before the crazy sergeant did worse. In that scenario, retreat would be more attractive to them than attack.

If violence is needed, you need to apply only as much as necessary to accomplish your objective. But have a backup in case it goes to the next level. Escalate slowly and leave room for retreat. Give them a chance to get back in line and follow orders. Stern warnings, physical pressure only if justified, violence only if necessary. I had wood in my right hand but steel in my left; I was willing to use both if needed, and it showed.

You might ask, what the hell does this have to do with Special Forces? But I was learning a bit about getting people to do what you wanted them to do where you didn't have a system backing you up, and you were at a disadvantage if violence erupted. Never under-estimate the value of a good bluff. And never bluff if you don't have a good backup in case it doesn't work. I would run into a situation like this again later.

Luckily for him (and me), he backed off when I stood up and hitting him wasn't necessary. I was assisted by the fact that my fellow Rangers had had their encounters earlier, and the reputation of Ranger NCOs had already been established. Ranger NCOs do not rely solely on the Uniform Code of Military Justice to enforce discipline. We use it when appropriate, but the bottom line is if we are in charge, we take charge, and anyone who doesn't like it will have an immediate problem on their hands, not a possible problem in six months.

Like I said, the officers and senior NCOs in the 509th were glad to see the Rangers. We had a few troublemakers, but they didn't run the unit. We did.

And it was a good unit. The majority of the troopers in the 509th were hard-charging soldiers who just wanted to do soldier stuff. But the limited training areas and resources made keeping busy difficult.

Don't get me wrong, it was great being in Vicenza. It was an old city with historic cathedrals, ancient cobblestone streets, and beautiful views. In the morning, you could run the steps up Monte Barico to the Basilica of Saint Mary and visit a variety of trattorias and small bars in the evening. The food was great, the people were wonderful, and the wine was incredible.

We could travel off-post to train in the Dolomite mountains and other training sites used by the Italian Army and conduct winter training on world-class ski slopes in the Italian Alps in the winter.

But that took money. When Jimmy Carter became President in 1977, he and Congress significantly cut military funding. At one point, the Army was so broke that it couldn't afford fuel for the trucks, gas for the aircraft, or ammunition for training, so off-post training opportunities started drying up.

The chain of command came up with a series of competitions to keep the boys busy. We had a 509th football team and a 509th parachute demo team, with roughly fifty guys who trained almost exclusively for those sports. They were exempt from other duties, so even though they might have been on your squad, they were rarely there for training and didn't perform other duties like guard duty or CQ.

When we went to the field, they went to their assigned unit. And they were in shape, even if their soldier skills were a bit rusty. I had two defensive and one offensive football lineman in my squad. The only time I saw them was on the annual ARTEP (Army Readiness Training and Evaluation Program), but when they did show up, I had one hell of a lead fire team in my squad.

In my last deployment with the 509th, we went to Wildflicken on an annual ARTEP. We conducted the largest air assault exercise in U.S. Army Europe history, replicating what the guys in Vietnam did on battalion-level air assaults. Many of our NCOs had been on those missions, and we used the same tactics.

We moved cross country on foot for twenty kilometers, about twelve and a half miles a day for three days. A twelve-mile road march is tough. Movement cross country through heavily wooded areas, across streams and fence lines for that distance is more than most light infantry units would even attempt. But the exercise planners were all mechanized officers, and their standard for movement each day in tracked vehicles was twenty kilometers, so we had to keep up. We were clearing the wood lines they couldn't maneuver through.

The 509th not only kept up, but we ran into the opposing force battalion headquarters, overran it before they realized they were in trouble, and captured their commander with our lead platoon.

The 509th were good soldiers and excelled at what they did when they had a chance.

But with the political situation in the U.S. and Italy, we didn't get as many chances to train as we would have liked. That was frustrating but it didn't have a major impact, and we stayed busy coming up with training we could do on post and in the barracks.

The problem was that many of the guys involved in the sports teams and special projects were NCOs and exempt from the duty roster. That didn't leave a lot of NCOs eligible for Sergeant of the Guard (SOG) duty, Staff duty NCO (SDNCO), and the Charge of Quarters (CQ) roster. More events popped up as the battalion commander looked for other events to keep the troops busy, but it pulled more and more NCOs off the duty roster.

Usually, a buck Sergeant (SGT) was on the CQ roster, but only Staff Sergeants (SSG) were on the battalion SDNCO roster. Sergeant First Class (SFC), three up and two down, and First Sergeants (1SG) didn't pull duty. SGTs also pulled the Sergeant of the Guard supervising the guard force out at our Ammunition Supply Point.

It got so bad that the chain of command decided that a SGT could be assigned SDNCO, CQ and SOG since we didn't have enough SSGs eligible for duty. A CQ would be excused from duty the following day because he couldn't sleep. However, a SDNCO could sleep since he was only there if needed and had to report for regular duty the next day. The SOG could also get some rest to come to work the next day. Weekday duty, weekend duty, and SOG's were all on different schedules.

So I found myself one week coming in on Wednesday for my regular job, staying that night as Sergeant of the Guard, regular duty the next day, weekday CQ on Thursday night, and weekday Staff Duty on Friday night, which meant I had to come in for the briefing that day. There was no point in returning home just to return a few hours later. I had weekend CQ on Saturday morning when I got off Friday night duty as Staff Duty NCO that ran all day Saturday and that night, then started Sunday Staff duty in the morning that ran all day and that night. And again, since I could sleep as the Staff Duty NCO, I was back at work on Monday morning. Either someone running the duty rosters had a sick sense of humor or I was extremely unlucky.

And on Monday, the battalion commander announced four more team competitions where participants would be exempt from the duty roster.

I surrendered and volunteered for the Para Cross team.

CHAPTER THREE

Para Cross

Para Cross was an annual military competition held in Germany, involving four events over two days. It begins with a five-kilometer cross-country run through the woods and fields wearing boots and uniform, carrying a rucksack with a twenty-pound sandbag and a pistol. You climb a twenty-foot ladder obstacle along the route, cross a one-rope bridge over a stream, and your run is timed. The fastest times win points for that event.

That's immediately followed by a pistol competition at ten-meter targets. Competitors receive points based on accuracy. That's followed by a one-hundred-meter swim in heavy German field uniform over one log obstacle and under another, four swimmers competing with one another, points awarded based on time.

The second day involves four rounds of parachute freefall accuracy. A ten-centimeter disk is placed in a gravel pit, and teams of four jump from one thousand meters (three thousand two hundred and eighty feet) on one pass. They try to touch the disk with their foot as the first point of contact with the ground.

The physical events are not easy, but it's largely a matter of being in shape. We trained for months on fitness and technique because we would compete against the best military athletes in Europe, such as the British SAS and German Combat Swimmers. Fitness alone wouldn't be enough. Some of those guys competed for their nation in the Olympics. Technique counts in the run and swim, accuracy with a pistol when tired from the run, and parachute accuracy.

This didn't have much to do with being a good infantry paratrooper, but it did give me a lot of experience working with some world-class Special Operations personnel and learning some new skills that would come in handy later. A big part of Special Forces involves getting to where the work will be done. Often you can fly there and land, then drive to the area where you will be working. But if its in an area where U.S. Special Forces is not welcome you have to be a little more creative about the arrival. Freefall parachute is one of the infiltration techniques used by Special Forces.

All Special Forces soldiers are airborne qualified using round parachutes. But in the '70s square canopies were becoming common. A round canopy will drift pretty much where the wind takes it, and it often takes you somewhere you don't want to go. Trees, bodies of water and electric power lines do not make for a good entrance. A square canopy can be controlled and flown like an aircraft to a pin-point landing. But it takes training and experience.

We would use relatively new StratoCloud ram air canopies for the competition and jump in four-man stacks on one pass. To obtain the proper dispersion of four guys jumping in one pass, we would need to freefall until everyone was in the air and pull the ripcords simultaneously. If done properly, you wound up with four guys staggered vertically and horizontally, and you could all fly to the target and land one at a time.

It's a carefully choreographed dance that involves exiting the aircraft, freefalling in staggered formation, simultaneous opening, flying to the target in formation, and landing one at a time on the same ten-centimeter disk. Everything has to go right for the team to score the maximum points.

In those days, we jumped the main canopy in a container on our back, with a reserve mounted on our front, a "belly mount." Both canopies were opened by pulling a steel ripcord that pulled pins, which allowed the container to open and release the chute.

When the main canopy doesn't open, it's necessary to "cut it away" by activating canopy release connectors mounted on your shoulder. It's not easy, and they both have to be activated simultaneously, or one could jam and you wouldn't be able to cut away fully. If that happens, the reserve can get tangled up in the main and it's over for you. If you can cut away you pull a second ripcord on the reserve mounted on your belly.

The training cycle starts with a static line where a nylon strap attached to the aircraft pulls the pins out of the container and opens your parachute. You pull a dummy ripcord that's not attached to anything to prove you can do it without flipping or rolling over in freefall. If not, your static line will at least open the container, and hopefully, you won't get wrapped up in the parachute's lines before it opens.

Once you can exit and stay stable, you work a progression up to terminal velocity. The first time was unnerving. Any slight movement of your hands, feet, arms, or legs will result in the movement of your entire body in freefall. And it gets more pronounced as you fall and pick up speed. The trick is to relax and let the wind push your body into a naturally stable position.

But relaxing while plummeting toward the ground at one hundred and twenty miles an hour is easier said than done. When you come off the static line, you try a "hop and pop," just jump out and pull, then a five-second delay, then ten, and you are clear for freefall if you can stay stable. It takes about twelve seconds to reach terminal velocity, as fast as you are going to fall.

Once everyone is trained to fall, pull, and fly the canopy, you get into "jumping stacks." The stack leader exits the aircraft and falls flat and stable, facing toward the target to keep himself oriented in freefall. That in itself is tricky since you have to control your body during exit. It's easier to face upwind on exit, in the direction the aircraft is flying. That puts you directly into a "relative wind," where your body pushes against the air directly in front of you.

Hold your hand out the window when driving in a car, and you will feel a relative wind. You can "grab air" like that to keep your body stable if you know what you are doing as soon as you exit. You are moving forward at the same speed as the aircraft, so you will experience relative wind for the first few seconds until you slowdown in forward speed and accelerate in downward speed.

But you have to face the target behind the aircraft for accuracy jumps in a stack of four. You are relatively close to the ground when you exit, and there isn't a lot of time to fly around and get oriented. Everyone needs to make a clean exit, stay oriented for a short time in free fall, open simultaneously and immediately fly in formation toward the target. You either need to exit and spin around to face the rear or jump towards the rear. The problem with jumping towards the rear is that the relative wind works against you. Your feet are lower than your head during exit. But to be stable against the relative wind, you need to form a cup shape, where your legs and arms are slightly to the rear of your body. If you form that cup facing forward, you're in a natural position to be immediately stable. But if you do that standing upright and facing the rear, the cup is inverted and you will immediately flip, either sideways or head over heels.

You don't have a lot of time in free fall, so it's better to exit facing the rear. That entails diving out and down, not jumping up and out, tucking your legs up a bit, and extending your arms forward a bit so your body is riding the relative wind in the cup shape facing to the rear. It's tricky, and it takes practice. It's harder in a helicopter since the rotor wash pushes the wind down while the relative wind from the forward speed pushes back. And it's different depending on the type of helicopter or airplane you are jumping out of. There are a lot of factors to consider, and you need to know them all from experience.

If all goes well you wind up with four jumpers under canopy, stacked up with about four hundred feet of altitude between each man, staggered toward the rear, and headed toward the disk. The formation flies toward the disk, passes it slightly to one side, then turns, and each man in sequence sets up on an approach and gets into his brakes to settle the canopy more or less straight down into the pit.

Remember that this is a Ram Air canopy with a forward speed of twenty miles an hour in still air. A Ram Air canopy is not round like a normal parachute. It's square, with two layers of material that are sewn together in the back and open in the front, with additional material sewn inside the two layers to form seven open ended cells.

The parachute functions like an airplane wing. You fly forward faster than you drift down, and the air is "rammed" into the open end of the cell to keep the shape of the wing and allow you to fly the parachute like an aircraft.

You cannot land this running downwind, or you will seriously injure or kill yourself. Think about jumping out of a car at twenty miles an hour. You do NOT want to try that. You also do not want to slow the parachute down too much. Too slow, and you don't have enough air pressure flowing into the front to keep the wing inflated, and it collapses into a wad of material and suspension lines. And down you come a lot faster than you want to.

Under ideal conditions with highly trained parachutists, all four will fly to the target and touch the disk for a perfect score. But conditions are never ideal, especially with the ram air canopies we had in 1978. These were "ring and rope" StratoClouds, generally regarded as the most accurate canopies available at the time, but new technology. We didn't have "sliders" at that time.

A slider is a piece of material with grommets at the four corners that the parachute suspension lines run through. The slider is next to the canopy at the top of the lines when the parachute is packed. When the parachute is released from the bag it's packed in, the canopy snaps open almost instantly. If it isn't slowed down, it will break the back of the jumper or blow out cells. It means going from one hundred twenty miles an hour to about two miles an hour in less than two seconds.

The slider catches air like a canopy and must be forced down the suspension lines. It slows down the opening but still slams your teeth into your chest. Sliders are simple and they work quickly and safely. But the slider was new, and we didn't have them, so we jumped ring and rope.

"Ring and rope" was an early reefing system intended to do the same thing as a slider, but it was a lot more complicated. Little one-inch rings are sewn into the bottom of the canopy all along the edges. A sixty-foot piece of braided nylon rope is run through the rings all the way around and up through the center of the canopy through brass grommets.

The ends of the rope are connected to a spring-loaded pilot chute, a smaller parachute packed inside the container. When the rip cord is pulled, it pops out into the wind causing the flaps of the container on your back to fly open.

That pilot chute would pull the two thirty-foot sections of rope out first, then the canopy, then the suspension lines, about another fifteen feet worth, and then the canopy would start opening. While all this is happening you are still in free fall, roughly three seconds' worth. While fifty feet of lines and rope are coming off your back, you fall another three to four hundred feet. Don't roll or spin or you will have a mess.

In theory, once all that was deployed, the canopy would slowly open, and you could take control. But it never worked that way.

Unless you were falling fast, the end cells would not open completely. You had to pull a red suspension line down to pull the rope through the canopy grommets and pull the pilot chute down to the top of the canopy, so it wasn't flopping around and getting tangled on things.

Sometimes, two end cells on one side of the canopy were not opened, so you would spin under a partially open canopy and descend quickly.

At this point, you would have two pieces of that long rope and the red line hanging down below your boots. Don't get tangled up in all the mess; you will have a serious problem. If you had five of the seven cells inflated, you were still falling too fast to land safely, and the chute couldn't be controlled so you needed to pump the rear risers to pump air into the deflated cells. That would usually pull the rope back up where it was supposed to be along the bottom of the fully inflated canopy unless you got the rope tangled around your boots.

In short, you were one busy paratrooper while all four of you drifted toward the ground and the target. You had to clear up your problem, and fast. If your buddy below you had a fully open canopy and you had closed cells that needed to be opened, you were falling faster than him, and soon you would be even or below him, screwing up the stack formation. That would mean a traffic jam at the target where everyone has to land on the same spot.

It takes practice, skill, and a bit of luck to do it right.

We would also do demonstration jumps for crowds at football games and at the ParaCross competition where accuracy stacks were not involved and we could exit the aircraft at a higher altitude, which made the deployment of the parachute a little easier to deal with. We would tuck up our arms and legs and go into a "no lift dive," where your fall rate went from one hundred and twenty miles an hour to one hundred and eighty miles an hour. When you are ready, you pull the ripcord, "get big," and spread eagle, grabbing as much air as possible to slow down. The increased fall rate tends to make the deployment system work how it was meant to.

The only problem with the higher altitude jumps was that we usually jumped with a heavy steel bracket on each boot, each one holding one or two military smoke grenades. So that pilot chute, thirty feet of rope, canopy, and suspension lines were pulling off your back close to that hot smoke grenade and steel bracket while you fell—more potential fun to deal with in the air.

Between ParaCross, ARTEPs in Germany, Platoon Confidence Training, and the Basic NCO School in Germany, I was away from home quite a bit, mainly in Germany. I was married now and I had a new son as well.

Much as I enjoyed Italy and the 509th when I was in Bad Tolz at the Platoon Confidence Training, I was deeply impressed by the area and the unit. Our ParaCross competition was in Germany, and we went to Bad Tolz for other parachute accuracy competitions. I wanted to go to Bad Tolz, so I submitted a request for an inter-theater transfer. I wasn't Special Forces qualified, but PCT was taking Ranger-qualified NCOs with Ranger Battalion experience, so I gave it a shot. One year later, it came back approved.

CHAPTER FOUR

PCT

Being on the team at Platoon Confidence Training was a dream assignment. Flint Kaserne was where the 1/10th Special Forces Group was based. During WWII it was the Waffen SS officer school, captured by General Patton. It remained the only Kaserne in Germany that had never been formally handed back to the German government. It was beautiful, nestled in a valley with the German Alps as a backdrop. The U.S. forces based there got along well with the local community, and training at Bad Tolz was as simple as drawing your weapon and walking out the back gate.

We had permission to range anywhere in the mountains we wanted to train. There were miles and miles of mountains and fir trees, rushing streams, two helicopters based on the Kaserne, multiple drop zones within walking distance, and live-fire ranges for whatever we needed to shoot. It was heaven for a soldier who enjoyed doing soldier stuff.

Bad Tolz was one of the best Kaserns in Germany, with plenty of housing available for families, small enough to not be crowded, big enough to have its own PX and Commissary, and it even had an indoor pool left over from the time when it was the Officer School for the Waffen SS.

PCT was run out of Camp Worden, a small temporary camp in the woods near the ranges. We had a cliff face where we could teach mountaineering, rifle and pistol ranges, and, best of all, a wide swath of the German Alps where we could roam and train anytime we wanted.

PCT was primarily about training platoons from all over Europe, but we also had a wartime mission since we occupied slots for Special Forces ODAs. But not being trained as Special Forces, there were limits on what we could do.

As Rangers, we understood patrolling and reconnaissance. Special Reconnaissance is a Special Forces mission in which small teams infiltrate one hundred miles or more behind enemy lines to observe and report on enemy movements and activity. We were not trained or equipped to operate independently that far back, but since we were in Europe, there was another mission we could perform.

At that time, the U.S. Army trained hard to repel an invasion of Western Europe by the Soviet Union. It was assumed that if the invasion came, the Soviets would make it to Bad Tolz and further.

Bad Tolz is less than one hundred miles from Czechoslovakia and just to the south of Fulda, Germany. The terrain between East Germany and Fulda is relatively flat, open, and ideal for tank warfare. The Soviets had massed a considerable amount of their armor forces in East Germany in a position to attack along the "Fulda Gap" between hilly terrain that would slow their movement. That led to Frankfurt and Rhein-Main, the major airports used by the U.S. forces to move troops and materials into Europe by air.

A secondary invasion route was through Austria along the Danube River valley. That would allow Soviet forces to enter West Germany from the south, passing near Bad Tolz.

The PCT cadre was tasked to train for Special Reconnaissance, not as a deep insertion, but as a stay-behind element to report on Russian invading forces. This was well within the capability of the Rangers with a bit more specialized training.

Part of the training involved using the old PRC 74B radio. That's a High-Frequency radio that can send signals hundreds of miles if the antenna is correctly "cut" and positioned, bouncing the radio signal between the ground and the atmosphere to its intended destination. It was big, heavy, and complicated, but it worked and was the best technology available back then.

Additional training we needed was Soviet vehicle and equipment recognition and advanced Escape and Evasion techniques unique to Europe. There were two schools available to us to attend for that: the Long-Range Reconnaissance school run by the Belgian and German Army, and the German Einzelkampfer school

(Single Fighter), also commonly referred to by U.S. troops as the German Ranger or Commando course.

Both are excellent training. The Long Range Reconnaissance School trains you to recognize all Russian vehicles quickly and accurately. But the German Ranger Course would be the most important preparation for the next jump in my career. It includes land navigation over long distances and it's a timed event. To pass you need to run, not walk, for most of the course and you need to have your land navigation skills honed to a fine point if you are going to succeed.

The first "march" is seven kilometers long. That's seven-kilometers straight line distance, and you rarely can go in a straight line. You are given a map for ten minutes. During those ten minutes, you must sketch key points on the first leg of your route: hills, roads, landmarks, and anything that will help you find your point at the end. Then they take away the map.

Hopefully you did a good job because in the end, you need to find a ring with small pieces of wood painted a specific color and with a number or letter. You recover one of the pieces and take it with you. Then, you open an envelope with the instructions on the route to your next point.

The instructions are to move on a specific azimuth for a specific distance, then change azimuth for another leg, and a final azimuth to your next ring of little pieces of wood. This time, they hint at where the ring will be. "Eine hutte nur ein weg" for instance. A hut near a path. Each leg of the journey will go right across a deep lake, down a steep hill, and back up another steep hill or through dense underbrush. You can go around, but you had better know how to get back on azimuth, or you will never find your point. The 'hut near a path" leads to a large open field with dozens of little huts near paths where hay is stored to dry for winter. You better be right on, or you won't make your time.

I failed that first run because I was overconfident. So did a lot of others. We got additional instruction and understood how serious they were about this, and after that, we did better. But each time, the course gets longer and harder. Towards the end, you cover seventy kilometers as a team, about forty-two miles, with river crossings, obstacle courses, rock climbing, and other tasks. You get very good at land navigation. Another skill that would come in handy later.

After completing those courses, we were assigned to participate in the annual Long Range Recon exercise run in Europe each year. It's a week-long exercise where four-man teams jump into a designated area, set up a hide position, and wait for a jeep to pass by with a code letter on the side.

When you spot the code letter, you open an envelope with that letter and take out a stack of photos of Russian military vehicles. Each one has a number below it. You identify the vehicles and send a message to the exercise HQs with the code names of the vehicles and the number on the HF radio.

Then, the fun begins. Because waiting in the woods are German and Belgian infantry units whose job is to find and capture you. They have radio direction-finding equipment, and the instant you begin transmitting, they will triangulate your position. They have helicopters, trucks, and tracking dogs to help with the search.

Your job now is to escape and move to the next hide site, where you will do it all again. And at the end, you turn in your rucksack and radio and move seventy kilometers cross country on escape and evasion. For this part of the exercise, the Germans and Belgians brought more troops to cover every road and patrol the woods. You will be moving for two or three days, trying to evade with no food or water other than what you can find along the way.

At the end, a truck picked us up for the ride back to the main base. After three days of evading the minute our rear ends hit the seats of that truck, we all went to sleep. Somewhere along the way I started dreaming about cars careening around corners, tires squealing and sliding as the car tilted to the side.

Except I wasn't dreaming. The truck driver was also asleep and started running off the road. The driver tried to get back on the road but overcorrected, and the truck started rolling over.

I was knocked out when my head hit the pavement, and I woke up much later with ambulances arriving and taking us all to the hospital. Seventeen stitches in the back of my head, six in my chin, my night vision devices were in pieces hanging on the cord around my neck, and my rifle was broken in half. But I was alive.

We had fun. We even made a balloon jump with the Belgians and got their wings awarded to us. I was a bit sore and had to pad my helmet with an abdominal bandage to protect the stitches, but I wasn't going to miss the chance to make a balloon jump.

It was a windy day, so they only let the balloon go up about four hundred feet above the ground. You could casually converse with the people on the ground at that elevation while you waited your turn to jump. Most of my conversation was with the team medic reminding me that he wouldn't sew me back up if I pulled out all those stitches in my head.

That ended my European adventure. I was nearing the end of my tour and enlistment, so I enlisted for Ft. Benning as a Ranger Instructor. I had just been selected for E7, Sergeant First Class, and everyone had to have a tour as an instructor as part of their career progression. PCT didn't count since it wasn't an accredited course, and I didn't want to be a drill sergeant, so RI seemed like a good compromise. From there, I could go to SF or return to the Ranger battalion as a platoon sergeant.

But it wasn't to be. When I checked into the Ranger Department, the Sergeant Major told me that since I was at Ft. Benning, I would be selected to be a drill sergeant, and RI wouldn't protect me from that assignment. It was cheaper for the Army to move people already on Benning over to Drill, even if it meant back-to-back instructor assignments.

In fact, according to the Sergeant Major, "you might as well buy a house. We're short of drill sergeants and it's a five-year stabilized tour. You will be an RI here for two years until you get picked up for drill, five years on the trail, so figure you have seven years locked in at Ft. Benning. You have eight years in right now. You will be fifteen after drill sergeant, so you might as well just stay here until you hit twenty. We will bring you back here for a couple of years. You can go to the airborne department for a few years and then retire."

Oh, hell no. I'm a field soldier. I don't mind doing my part as an instructor for a few years, but not my whole career.

I wanted to be Special Forces. But I couldn't get there from where I was. The Army leadership hated SF and openly discussed getting rid of it altogether. No one would approve a transfer off Ft Benning for the next seven years.

So, I committed the ultimate betrayal of the NCO Corps by submitting my paperwork for the only thing that took precedence over drill sergeant at the time.

Officer Candidate School.

CHAPTER FIVE

Escape to Ft. Bragg

Most people spend their entire lives in one pursuit, sometimes regretting that, wondering what could have been. My life has been about radical change and reinventing myself into something completely new again and again. There isn't much I haven't tried at one time or another.

I left the NCO Corps and volunteered for OCS to become a mere Second Lieutenant. Basically, start over on a new career path. You might say it was the same; after all, I was still in the Army, and after I graduated from OCS, I would still be in the infantry, but the job of an officer is different from that of an NCO.

NCOs are soldiers, and they train other soldiers. The NCO corps does the hard day-to-day work of the Army. They are experts in weapons, field craft, and soldiers' leadership in harsh and dangerous conditions. It was what I signed up for.

Officers get some of that, to be sure, but they are groomed and prepared for promotion to General from day one. They get just enough time in the field to see if they can perform at that level, and then they are promoted to the next level or separated from service. It's very much up or out. Officers plan the next step; strategy, tactics, and logistics, and NCOs train and supervise the soldiers to carry out the plan that's on the table now.

With ten years in service, more like eleven before I would become a fully qualified officer and complete training, I would probably never become a General. That meant I would be shuffled to the side for promotions no matter how well I did, with the expectation that I would make Major and retire around twenty years in service.

Which suited me just fine. That meant there would be no rush to move me from my slot. I would get jobs that were higher risk, that "fast movers" avoided to protect their chance of promotion. OCS officers who graduated with NCO time tend to get multiple commands they don't need for promotion and get assignments that are difficult and off the beaten path. I reinvented myself as an officer and took what was viewed as a step down that allowed me to keep doing what I loved. It kept me working in the field with soldiers at the company level.

This began a whole new chapter on the road to Special Forces. It was not what I had planned, but there was an open path. But to get there, I had to go through OCS as an E7, sit through the same classes I had just been through during three months in the Advanced NCO Course, and sit through the same classes a third time in the Basic Infantry Officer Course for another three months. When I got to the Infantry Officers Course, I was shouting out the punch line for the instructor's jokes before they could get to it.

We were all glad to see me leave Ft Benning.

While in OCS, I was asked to fill out a "wish list" for my branch and first assignment. I put down the Infantry and the 82nd Airborne Division. Getting to Ft Bragg was my best bet for joining the Special Forces program. But I had a long way to go before that happened.

At that time, SF was not accepting 2nd Lieutenants. You needed at least one successful assignment as an officer before applying. Officers should have some proven experience before going to an A-Team. I got a rifle platoon in the 2/325th Parachute Regiment, one of the three infantry brigades in the 82nd Airborne Division. I spent one year there, made 1st Lieutenant, and was assigned to the Scout Platoon as platoon leader.

Transitioning to officer was pretty simple for me. I knew what a platoon sergeant should do and what all NCOs wished a platoon leader would do, and I had a good platoon sergeant in the 82nd. I stayed out of his way while he did his job and set out to find ways for my platoon to get better training opportunities.

Career-wise, things were going well. On the home front, not so much. I liked the new life, but my wife did not. NCO social life is rough and tumble, but officer social life tends to be more structured and formal. The officers expect their wives to participate and support the family support group while we are deployed, and so on. My wife was having none of it. We gradually drifted apart, and divorce happened right as my time as a scout platoon leader was drawing to a close.

I didn't want a divorce, but I didn't have much choice in the matter. She got the house and our son and temporary alimony plus child support. I got one of the two cars and all the bills, as is usual in these things. It hurt, especially being separated from my son, but it was what it was. I fought for custody in court, but the man never won in North Carolina in a custody dispute.

What broke my heart was the impact it was having on my son. Both sides are responsible for what happens in a divorce, and plenty of the blame lies with me, but I didn't want a divorce. But my wife had arranged things to ensure she had control of the situation, so it was happening whether I wanted it or not.

It was a good time to go forward with SF since I would be single anyway. SF has a high divorce rate, largely due to the long separations during back-to-back deployments, but even that was denied to me. As an officer, I was required to have a college degree, and while I had two years' worth of college credits, that wasn't a degree. I was told I needed to go to degree completion for a year and a half at Army expense, or I would not be promoted to captain when I was eligible, and I would be separated from service shortly after that.

Faced with the loss of my family and a delay of another year and a half before I could go to SF, things were looking pretty bleak.

I was depressed and for a while nothing seemed important anymore. I hadn't done freefall parachuting for a long time, so I headed to the sports parachute club and asked if I could jump my rig. I was an experienced skydiver with over three hundred free falls and a C license, but I hadn't jumped in a long time. A little free fall might clear my head and raise my spirits.

But since I hadn't been in the air for a while, I was required to do a refresher progression. This involves some practice on the ground for emergency procedures, mainly cutting away from your main canopy and activating your reserve. If your main canopy doesn't open or you can't control it, you cut it away by activating the canopy releases on your shoulders.

Before jumping my rig, I needed to do a static line jump with a student freefall rig. The static line would deploy my parachute, but I would make a free fall exit of the aircraft, arms and legs extended to demonstrate I could make a stable free fall exit and pull a simulated rip cord.

Then I could do a "hop and pop," jumping out and pulling a real rip cord, followed by a jump where I would delay pulling for five seconds, then ten. If I could do that, I could handle the exit and terminal velocity, where you are falling as fast as you will fall and small movements of your body cause major changes.

They gave me a student rig, and as they checked it before the jump, the jump master noticed the static line wasn't very secure in the closure loop. The static line was attached to the pilot chute of a round free-fall canopy with a piece of cotton string that would break free when the canopy was fully extended. The container on my back that held the parachute was four flaps of material held closed by brass grommets with a loop of nylon suspension line routed through the grommets. The suspension line loop was connected to one flap by a knot at the end with a metal washer to prevent the knot from being pulled through the grommet. The loop was routed through the grommets on all four flaps. A section of the static line was folded, and the static line loop was inserted into the suspension line loop, holding the whole package closed.

When the falling jumper elongated the static line, it would pull itself out of the loop of the suspension line, elongating the parachute and breaking away, allowing the parachute to open.

The knotted loop with the washer had fallen out at some point and had been replaced. But the new loop was too long and didn't hold the static line firmly in place. The loose loop of the suspension line allowed the static line to fall out and might allow the pilot chute to pop out inside the aircraft.

The jump master made a couple of overhand knots in the suspension line loops to shorten it up. But he made the knots on the wrong side of the metal washer, which would allow the knots to be pulled through the grommets.

I made a normal exit at three thousand feet and pulled my dummy ripcord, waiting for the jerk of the opening parachute. But all I felt was a slight pop, and I kept falling.

Remember the first chapter of this book? At that moment, I had less than twenty seconds to live.

Twelve seconds to reach terminal velocity.

Eight and a half seconds later, I would impact the ground at terminal velocity.

Four seconds had already elapsed while I waited for the static line to open my parachute. It wasn't open, and I knew I had a problem.

Sixteen seconds remaining.

I rotated my head to look over my shoulder and saw flailing crown lines from the top of the parachute, lots of parachute material, and suspension lines. The static line had yanked the knots through all the grommets, locking the container shut, but the portion tied to the crown lines pulled them out through the flaps of the still-closed container. About half the canopy followed, but the pilot chute was still locked inside.

In free-fall terminology, this is called a horseshoe malfunction, and it is one of the most serious malfunctions. People often die in this situation. If you cut the canopy away, it will just trail above you, still attached to your harness where it is fouled.

If you pull your reserve with all that material flapping around above you, there is a good chance that the reserve pilot chute will also get caught in the main canopy. Then you're dead.

That took two seconds. Fourteen seconds remaining.

I looked at it for another two seconds, thinking maybe it was for the best. I had lost my marriage and my son, and I didn't see much point in going on anyway; this was an easy way out. But that attitude didn't stick. Rangers don't give up.

Twelve seconds remaining. Nearing terminal velocity.

I rolled onto my back to give myself room to wrap up as much material as possible with my left arm. Then, I rolled back so I was face-to-earth and sat up in free fall, trying to pull as much material down to me as I could. Two more seconds.

Ten seconds remaining. Almost at terminal velocity. The ground was coming up to meet me.

I arched my back, pushing my chest as far up as I could, and fired my chest-mounted reserve with my right hand. The pilot chute deployed cleanly, and I was under a good reserve canopy, with the main flapping around my body and legs. Two more seconds elapsed.

It took a lot less time to do than it took you to read about it. But I had been in freefall for several seconds when I was supposed to be under an open canopy almost immediately. But I was a lot lower now and well off the drop zone.

I looked down to see where I was and saw rows of little square boxes with green dots inside them. This looked familiar, but I couldn't figure out what I was looking at.

Then I heard a deep, booming voice coming from behind me. "Is there anyone down range? Is there anyone downrange? Is there anyone downrange? Firers, watch your lanes!"

And the little green dots began popping into an upright position for the soldiers on the firing line behind me to start shooting.

I was over a rifle qualification range, and no one had noticed the parachute drifting down onto the target array.

I wasn't in any danger other than an occasional ricochet coming up, but it was still not good. And I certainly didn't want to land down there. I grabbed suspension lines and spilled air out one side of the canopy, causing it to slip off and head toward the wood line outside the impact area.

I made it out of the impact area, but now I was over tall pine trees, and there was nowhere to land but in the trees. I got into the tree landing position, legs together, arms crossed, and hands covering the armpits to help prevent a broken tree branch from spearing me in the armpit and severing the big artery there.

I hit the top of a twenty-foot pine tree, broke a branch with my feet, and fell through. As the canopy entered the tree, it collapsed, increasing my velocity toward the ground.

And just before I hit the ground, the canopy caught the tree top and snagged. The tree bowed, my boots gently touched the ground, and I stopped, suspended a few inches off the deck.

I got out of my parachute and waved to the helicopter that was now hovering overhead, trying to see if I was OK.

When I returned to the DZ, I approached the now very apologetic jump master who had almost killed me and asked him a question.

"Look, I know I'm supposed to do a hop and pop, a five-second and a ten-second delay next. But considering I did all those in this one jump, plus a reserve activation under a horseshoe malfunction and a tree landing, can I jump my rig now?"

"No problem, do whatever you want to do. You're good to go!"

That cleared my head and gave me some direction. Life goes on, and although I missed my son and continued to worry about him, I had to get my act together and find out what life had in store for me.

As an officer, I was required to have a university degree. When I went to OCS, I had about two years of credits, one year from North Georgia College and the rest from various other universities. Like many NCOs, I had been taking college courses at night and on weekends, working toward my degree.

After your initial assignment as an officer, the Army wants you to finish that degree. They give you a year and a half to do it full-time. Ft. Bragg had cut a deal with Campbell University for officers needing a degree for free, provided they signed over their

GI Bill payments to the university. That was the best deal around at the time.

I headed off to Campbell University to complete my degree. Probably the best thing that could have happened to me at that time. No money with alimony and child support payments, but I didn't need much; studying took up all my time. University wasn't nearly as hard as I remembered, and I enjoyed my time there.

Campbell was and is a Baptist university, and it emphasizes its brand of religion in its curriculum. Nothing was wrong with that, but I didn't agree with all of their closely held beliefs, which would sometimes create a little friction. I was enrolling when the first sign of friction popped up.

"Welcome, Mr. Brewer; here is your schedule. I need you to fill out these forms and sign this declaration. I also need you to fill this out for your dormitory room. You are a junior, so you will share a room until you reach senior status."

I looked over the declaration that says I promise I will not consume alcoholic beverages on or off campus as long as I am enrolled in the university. OK, this is getting silly.

"Ma'am, I think there is some mistake here. I am a degree completion, active-duty military officer. I'm well over thirty. I will not be staying in your dormitory, and I have no intention of signing this declaration."

"Well, Mr. Brewer, that is our policy, and if you want to go to school here, you WILL live in our dormitory, and you WILL not drink alcohol. Do you understand?"

"Lady, I understand that this university signed an agreement with the US Army to provide us an education, and no one said we had to convert to Baptist. I enjoy a beer occasionally; in fact, I feel like having one right now after listening to you. And you don't want me living in your dorm with your baby Baptists; I promise you I will be a bad influence."

That took a trip to the Dean's office to defuse, but at least everyone agreed that I would be a bad influence, and I was allowed to live off campus. Wise move, turning a newly divorced thirty-two-year-old loose among the coeds would not have been good for anyone. But it brought home another aspect of Special Forces I had not yet fully learned; accepting aspects of another groups culture you don't agree with but you must respect without compromising your own values.

My only real challenge in the institute of higher learning was statistics. Our professor looked like the creature from the old TV show "Tales of the Crypt" but without the personality. He would talk in this monotone voice, putting you to sleep in about five minutes. On top of that the subject matter was not exactly thrilling.

In his monotone voice he slowly told the class that most of us would fail the course, but he would try his best to teach us something. Really encouraging, professor. I'm on degree completion and cannot afford to fail. I can't afford a D, that could put me out of the program.

Worse, he insisted on using an old textbook full of errors. He would lecture for an hour and a half, covering all four blackboards with equations that had nothing to do with the last assignment we hoped he would explain. I would do all the exercises he assigned, but there was always one that just didn't come out with the answer in the back of the book, no matter how many times I would work it.

After an hour and a half of watching him explain some obscure equation, he would finally announce, "And I'm sure you realize by now this has nothing to do with your assignment, but it is how we predicted bean consumption of soldiers in World War II, and I thought you might find it interesting."

I would ask about the problem in the assignment, and he would cackle, "Yes, Mr. Brewer, very good. The answer in the back is incorrect. If you couldn't come up with that answer, you might have completed the equation correctly."

Might have? Give me a break.

He only gave four exams during the entire semester. I took his first test, and it looked worse than the textbook. I could barely understand half of the test. I got my results, and I scored 40 percent.

An F.

This isn't happening to me.

I needed to ace the remaining tests to keep my grade point average. I was an A student now, but I wouldn't be much longer if I couldn't change the outcome of the next test.

I studied as hard as possible, worked on every problem, and took the second test.

It came back a 50.

Another F.

I needed 100 percent on the last two tests just to get a C.

I hired a tutor who had taken his class and got a lot of insight into how he set his tests up. This guy loves to mess with his students, but I see a method to his madness, and I am more confident going into test number three.

But as I'm taking the test, there are at least one-third of the problems where I cannot understand what he is trying to ask us. Finally, I've had it. I get up from my desk and carry my test up to him.

"Mr. Brewer, I have not called in time, and you could not have finished your exam. Please return to your seat."

"Oh, I'm done all right. I have done everything I can to absorb this subject, but you are the worst excuse for an instructor I have ever had the misfortune to suffer under. You couldn't teach someone to pour water out of a boot without the instructions on the bottom. This test is a joke, and apparently the only reason you are running this course is to see how many students you can fail. Well, you win. I'm dropping the course before you fail me." And I tear up the test and throw it on his desk.

I storm out and try to shake it off outside. A little later, two of my classmates came up to me. "I can't believe you just did that."

"Yeah, I know, that was a little over the top. But that guy was making me crazy, and I need this course to graduate. I don't know what I'm going to do now. There is no way I would get better than 70 percent on that test, and I needed to max it to make up for the 40 on the first test and the 50 on the second."

My young classmate's jaw drops at that, and she says, "You have the highest grade point average of anyone who has ever taken the class!" Don't you know he grades on a curve? No one has ever actually passed one of his tests; you would have been the first one ever to do so if you had turned it in!"

What's that thing Homer Simpson says at moments like this? Duogh!

I returned to his classroom, hat in hand, to apologize and beg forgiveness. I really do need this class.

"Yes, Mr. Brewer, I was quite surprised by your little show. Your diligence is paying off, and you have some marginal promise as a future statistician. However, you have already walked out on this test, so I am obligated to give you a zero. But if you can match your performance on the next test with the curve, you can get a C for the course."

Duogh!

I made it through and didn't get into much more trouble, although the friction would continue. More lessons learned for the future in SF; those who would teach you will not always tell you what it is you are supposed to learn. They create situations where you must figure out the right path forward on your own not knowing what the acceptable standard is. Try as hard as you can and don't give up. And most important, if you fall, pick yourself up and keep going.

Next, I was off to the Officer Advanced Course at Ft. Benning. I was promoted to Captain while I was at Campbell U., and I was being prepared to command an infantry company. The Army likes to see light infantry platoon leaders take command of mechanized infantry companies as their next step, so they learn all aspects of infantry combat. But I had other plans.

While at the Advanced Course I applied for Special Forces training. SF was still the red-headed stepchild of the Army and the Department of the Army wanted to get rid of SF completely, but SF had some powerful allies in Congress who were not allowing that to happen.

Volunteering for SF didn't make me popular at the Infantry School. As one senior officer explained, "You must understand, SF is obsolete and we are getting rid of it. By going there you are a traitor to the Infantry. We won't take you back when we get rid of SF. Your career is basically over."

Maybe so, but I liked the little taste of SF I received in Bad Tolz and wanted more. I'll take my chances. After six months at Ft. Benning, I was again back at Ft. Bragg in the Special Forces Qualification Course holding company.

My marriage was over and my NCO career was behind me. I had been successful as an infantry officer but left that behind me, taking a big chance to get through one of the toughest courses that most who attempt do not pass.

And I was thirty-four years old in a course many twenty-year-olds can't handle.

Maybe I was crazy. If I failed, there would be no coming back. I had fourteen years in the Army at this point and failing meant losing the twenty-year pension. I had already done all the hard physical things expected of me as an infantry officer. I could have taken an infantry command in a mechanized infantry unit and then maybe go back to the Ranger battalion as a captain.

But I had made my decision, and I was taking the challenge.

CHAPTER SIX

The Q Course

This was 1988, and many things changed for the Army and Special Forces. The Embassy hostage rescue in Iran had failed very publicly, and Congress was not happy. Special Operations were blamed but the real problem was the usual bickering between the services.

There wasn't a truly joint command controlling all the training and combat services. Everyone had competing and separate budgets. As a result, Army radios couldn't talk to Air Force aircraft, the Navy didn't spend much money on helicopters that primarily supported infantry, and no one was used to working with other services.

This was the primary reason for the failure of the hostage rescue mission. Again, we had serious problems with this in Grenada.

This forced the creation of the U.S. Special Operations Command, or USSOCOM, directed by Congress. It also explained why the regular Army despised SF. The need was recognized after the failure of the Iranian hostage rescue, reinforced by the mishandling of Special Operations Forces in the Grenada invasion. Still, it took five years and an act of Congress to force the solution down the throat of the big Army.

USSOCOM was the only truly joint training and combat command, and they controlled Army Special Forces, the Ranger Battalions, Army Special Operations helicopters, Navy SEALs,

Air Force Combat Controllers, and Air Force Special Operations aircraft.

The theater CINCs still had combat control of Special Operations Forces assigned to their theater. Still, the president could choose to send Special Operations into a theater under the direct control of USSOCOM and their subordinate commands.

The services didn't like giving up forces to this new command; they didn't like that they didn't have control of them in combat, and they really didn't like that USSOCOM had their own budget, and a pretty large one at that.

The same issues that had forced the creation of the Air Force from the Army Air Corps had nearly created yet another service. But the decision was made to leave the various units under their parent service but with operational and budget control under USSOCOM. This left a lot to be sorted out.

So we were truly on a one-way street, and there was no turning back from here. Failure of the course would not just mean going back to the Infantry; it would mean going back as a despised traitor to the Infantry who wasn't good enough for SF.

Motivation to succeed was high.

And SF wasn't about to make it easy. SF in the old days was an assignment, and we had infantry, military intelligence, and even armor and artillery officers commanding A teams and companies. You served a few years in SF, returned to your parent branch, and returned to SF later if that was what you wanted. But it was off your career track, and spending too much time in SF meant there would be no promotion in your branch.

So USSOCOM had decided to make SF a separate Army branch. They resurrected the crossed arrows from the days of the Special Service Force in World War II, and once you were in SF, you stayed there for your entire career.

They also changed the qualification course. A Special Forces Lieutenant Colonel came in to brief the officers on the course and explain it to us.

"You will not be going through the course I went through. Your course will be harder, a lot harder. We will try many new things in this course; some will work, and some won't. After your course, we will begin something we call Selection and Assessment. You will get the events for that, but they will be integrated into your Phase 1 course. We will decide if it's too hard and needs to be modified or made harder. It will be tough on you, but we need tough officers for the new branch, and you will be the first to go through this program. I wish you good luck and hope to see you graduate."

Once again, I managed to get myself into the toughest thing I could find without realizing what I had volunteered for. And this time, I was coming in at thirty-four, the second oldest man in the class. It is what it is, as they say.

The PT test for entry was on the 18-21 year old standard. The Army had separate scoring tables for men and women and those tables had separate scores based on age. The toughest standard was 18-21 men. No slack, 240 out of 300, or go home. OK, no problem. I was prepared for that. The cadre in the reception and holding company ran the test and daily PT to get everyone in the best shape possible (or at least weed out those who wouldn't last), and once we had gathered enough people to form a class, we headed out to Camp Mackall for Phase 1 training.

Phase 1 in the old days was equal parts "How bad do you want this?" gut check events, light infantry tactics, land navigation, and other skills common to all team members. Lots of PT, road marches, the "Nasty Nick" obstacle course, classroom instruction, patrolling, and the land navigation course.

Physical Training, or PT, in August at Ft. Bragg is hot and miserable. Staying hydrated was key to survival. It consists of long road marches, pushing a jeep trailer loaded with sandbags and missing one wheel, carrying two twenty-pound water cans plus your sixty-five-pound rucksack and weapon on road marches and other fun stuff. All were part of the curriculum and designed to wear you down, make you quit, or slack off on your buddies.

Everyone was watched and evaluated to see how well they worked as part of a team under extreme stress. The class of three hundred and sixty-five students began melting away like snow in the hot North Carolina sun.

The "Nasty Nick" obstacle course was a major gate you had to pass through. Log obstacles you had to climb up that towered a good fifty feet off the ground, then walk across logs spaced two feet apart with nothing but a long way to the ground between. Obstacles that require strength to get up and dexterity to get around. Today, everyone wears a Pro-Tec helmet on the course; I suspect it is because someone died from falling off the taller obstacles. Back then, occasional serious injuries were expected and considered part of the course. Your buddy just fell and broke his arm on the obstacle, and you are next one up. Just how bad do you want this?

One obstacle was a thirty-six-inch concrete drainpipe that ran a couple of hundred meters underground with several turns along the way. After the first few meters, it was dark the rest of the way. With men in front of you and behind you, it was stifling hot, tight, and claustrophobic. But there was also no way out once you started other than forward.

The man behind me lost it in the dark about halfway through and couldn't keep going. The event is timed, and if you don't push hard, you won't pass, so there was no time to back out and let him out, and in any case, he was in a panic and couldn't move.

I took my patrol cap with the Ranger eye tape that glows in the dark and stuck it in my belt so it hung down over my rear end. "Can you see that buddy?"

"Yeah, I see it."

"Then follow it."

And I took off. After a second or two, he followed, breathing hard but moving.

We got an introduction to some skills we would need as SF soldiers. One was communications. In those days, a good SF communicator could send Morse Code messages a thousand miles using a car battery, a barbed wire fence, and a few pieces of electronics you could pick up in any electronics store. Of course, It works much better with a radio, but it's technically possible.

The communications instructor explained how radio wave propagation works when one student stood up in the class and said, "That's not exactly correct, sergeant."

"Really. Then how about you explain how it works to the class?"

The student then launched into a ten-minute dissertation on radio wave propagation, atmospherics and frequencies, using words none of us understood and terms that had the instructor's jaw-dropping.

The instructor looked at the student and asked him, "What the hell are you, some kind of rocket scientist?"

"Well, actually, sergeant, I am involved in the solid fuel propulsion systems project at the Marshall Space Flight Center, but my hobby is high-frequency radio. I'm a 20th Special Forces Group National Guardsman here for the Q course before returning to my regular job.

You meet the most interesting people here.

Most SF training is team-focused, but land navigation is individual. You are warned up front not to speak to anyone you see on the course, not to help anyone, and stay off the roads. Any of these offenses can get you dropped from the course.

The course is called the "Star Course." It is a series of points you need to reach and check in with an instructor before the sun comes up. You start at sunset. The points are laid out so you cross back and forth across the area, forming a star pattern. Legs of the course range from two or three to nine kilometers. If you are not moving fast and know what you are doing, you won't make it before the sun comes up.

And it's one hell of a course. It's got it all: hills, thick scrub oak thickets, deep ravines filled with thorny vines, swamp, it's all there. They must have looked for a long time to find a place that was that miserable.

Our graded course began at sundown. I pulled out my map to plot my first point, and a large raindrop plopped onto the map, followed by more. It would pour rain for the next three days and nights.

And there were traps along the way. Numerous instructors ran the course in the same uniform we were in, imitating students and begging for help locating their points. If you helped them, you would have been out of the course.

That's one of the toughest land nav courses I had run, and I do mean run, but the German Ranger school I attended when I was an NCO in the 10th Special Forces Group in Bad Tolz had prepared me for that. While challenging, I hit my last point right as the sky was getting light on the horizon, but before dawn. It had rained hard all night, and it was still coming down.

After a few hours, we had most of the class gathered up. The instructors called out all the "first-time go's" to get on the trucks and head back to the camp. If you failed the course, you got another chance but had to go right back out that night after sleeping in the rain in the field. No hot chow or coffee for re-tests.

I climbed on the truck along with fourteen other guys and waited. The lead instructor shouted, "I won't say it again; if you are a first-time go, get on the truck now, or you can run the course again!"

The class leader stepped out of the crowd of wet, cold, and miserable students and replied in an unhappy voice, "Those <u>are</u> all the first-time go's sergeant."

A moment of shocked silence; then the truck headed back to the rear. They would run the course again, not once, but twice, just to get enough people through to make a class. We started with almost four hundred students but would leave Phase 1 with just over one hundred.

After Phase 1 the officers went straight to SERE school. Another one of the toughest courses in the Army. It wasn't part of the Q course before, but it was now for the officers. It teaches survival skills and eating what you can find in the forest, followed by a week in the Resistance Laboratory.

The survival part is first rate. One class in particular stands out in my memory. We were driven out to a place in the North Carolina woods next to an abandoned water cistern with what looked like an old redneck standing by in overalls and a Caterpillar tractor hat, wearing a black eye patch like a pirate and a big grin. Lesson number one is never to underestimate a person based solely on appearances.

"Boys" he said, "My name is Sergeant Major Nail, US Army Special Forces, retired. I'm gonna be your instructor for the next couple of days. We have a lot to cover, but first, I like to do a little gain attention step to see if you are serious about training. If you want to hear what I have to say, follow me."

The wall of the water cistern behind him had one side which had collapsed into a pile of rubble. But three-quarters of it still rose a good twenty feet into the air. The ground was littered with broken concrete pieces in the center of the cistern and around the sides. The broken edge of the cistern was a tangle of broken concrete and steel rebar that sloped rapidly up to the rim and was about six inches wide.

With that short introduction, that old one-eyed man scampered up that tangle of concrete and rebar like a mountain goat and stood on the rim, grinning at us.

"Come on, boys, this is required training, and if you can't do it, you might as well just walk back to Camp Mackall and turn in your gear."

After a moment of hesitation, we climb up onto the rim and carefully walk around the top, one continuous line of wanna-be Special Forces officers. One falls, breaking his arm on the rubble below, and must be medevac'd out. His course is over.

"That's about right," says the Sergeant Major, nodding and chewing his Red Man. "We usually lose at least one in this class who isn't up to it; good to get it out of the way right up front."

OK, he definitely has our attention now.

We spent two days with the Sergeant Major, learning a long list of invaluable techniques and "tricks of the trade," as he puts it. Someone used his head when he set up this course, and there is a wealth of experience we will be exposed to here. Old retired Special Forces officers and sergeants have been hired to teach new students what they learned the hard way, with over thirty years of experience in combat.

For instance, how to climb that twenty-foot sheer wall of the cistern. He directed us to get up the undamaged side any way we could. After trying to build human pyramids and other techniques that didn't work, he picked up a six-foot long two by four-inch piece of pine wood, tied a rope around one end, and threw the board over the wall.

The long end of the board hung on the other side of the wall, with the rope trailing back to our side. The sergeant major told one of our smaller students, "Grab that rope, boy, and get up that wall. Stay close to the wall; the board on the other side will act like a lever and hold you up. Your weight isn't enough pulling on that short end to let the long end come up."

And lo and behold, he was right. Once up, the smaller man put his boot on the side of the board, and even the larger guys could climb the rope. Simple physics applied to combat tasks.

The sergeant major culminated his training by showing us how the Viet Cong had infiltrated heavily defended US bases. We were taught to prepare wooden supports that looked a lot like a slingshot, just a Y-shaped piece of wood about a foot long.

Every man carried two of these. The sergeant major then took us to his training area, a section of barriers consisting of rolls of concertina wire staked down to the ground and piled high, one on top of the other forming a pyramid. Concertina razor wire will cut you to ribbons if you push through it, tangling you in the coils.

In front of the concertina was Tanglefoot. That's row upon row of single strands of barbed wire attached to stakes six inches from the ground. The wire runs left to right and diagonally forms a lattice. It's too low to crawl under, and if you try to run through it, your boots slip through the lattice, and the wire catches your leg.

This had trip wires running through it tied to small explosive charges and flares. The good sergeant major took up position in a guard tower overlooking all this with a searchlight that he would traverse back and forth while we tried to get through the wire after dark.

We were taught to strip down so our clothes wouldn't snag in the wire and camouflage our bodies to look like the dirt under the wire. Then we would take a weapon and a satchel charge each and slip under the tanglefoot wire. When the searchlight moved to another section, we would move, freezing in place when the light came back our way.

The lead man would look for trip wires that ran the length of the barbed wire between the stakes. He would trace the wire back to its source. Some trip wires can be cut, some are spring-loaded and will fire if cut, and they need to be disarmed at the source with a safety wire.

Once the trip wires were removed the lead man would take one of the Y-shaped sticks and rotate it up under the wire, Y-shape up. This would push the wire up another six inches. It is not enough to attract attention, but just enough to crawl under the wire slowly with his head turned to the side.

We would follow in a trail formation, the last man taking down the Y-shaped sticks and passing them up to the front until we were on the other side of the wire. Amazing training from an amazing man; it was a privilege to meet him, let alone learn from him.

After the survival skills portion of SERE, we prepared for the resistance lab. The resistance lab is a prisoner-of-war compound where you practice resistance to interrogation and surviving capture and imprisonment under brutal conditions.

Ranger SERE back in the mid-70s had been a good introduction to this, but this course was light years beyond what we had in the Ranger Course. They trick you into compromising information or betraying your buddies and are experts at finding and exploiting your vulnerable areas.

They train you to resist, recover, and keep going when you fail. You will fail at some point in that course. It's part of the course. They will break you, trick you, find a way to get you to compromise your values, or at least appear that you did in front of your buddies. The point is anyone can be broken, but resist as much as you can, shake it off, get back on your feet, and keep resisting if you make a mistake.

The training leaves a mark on you. No one goes through that course without it changing them in some way. There are psychiatric counselors at the end to help ensure a positive outcome, but not everyone can take it. Once again, we lose a few who are just not cut out for this and a few who are too injured to continue.

But again, we have a wealth of experience to learn from. Before we go into the resistance lab, we have the opportunity to talk to men who have experienced prisoner-of-war camps firsthand. Nick Rowe's buddy Dan Pistner is there. Nick Rowe suffered in a Vietnamese prisoner-of-war camp for five years, and Dan was there with him for much of that time before he was released. We learn how to survive under those conditions from men who have endured them.

There is one older gentleman in particular who was in a German prisoner-of-war camp during World War 2.

He had escaped three times, once staying free for almost a year before being recaptured, trying to make his way out of Germany and back to allied lines. "You must speak German pretty well," I asked.

"No, not at all." He replied. "In fact, that's how I got caught. I was asleep in a train station waiting to catch a train, and a Gestapo officer told me to get up and move on. I woke up and asked him what time it was. The only problem was I asked in English."

He went on to explain that at that time Germany was losing the war, and Allied aircraft were bombing German cities every day. He would hide in the woods until he heard bombing, then come out and go into the city.

There would be assistance for the survivors set up by the German Army, with soup lines, new clothes, and blankets, and anyone who wanted to leave the city could get on the train and go anywhere in Germany where they had relatives or someone who would take them in.

Everyone was head down and miserable and usually no one spoke to or noticed the single man in line with everyone else. He just got in line, kept his mouth shut, got a meal, and rode on toward the border.

Invaluable training from people who had been there, I have never seen it done better before or since.

After SERE, we go into the SF Officers training for Phase 2. Again, much of this is classified, but it is still intense, and it covers a broad spectrum of what you will need to lead an A-Team. A lot of the course is devoted to portions of what the enlisted members of the class are going through, both to give you some basic skills in each Special Forces specialty and to let you know what your NCOs are capable of doing. Weapons, explosives, medical, and communications are all on the schedule.

It culminates with Phase 3, Operation Robin Sage, the final exercise. This is where the enlisted and the officers form A teams, are assigned missions and plan in isolation, then infiltrate the training area and link up with guerrilla forces that you train, organize, and lead in missions.

We meet our teammates who have completed their specialty courses and share stories about what we learned. At first we are back at Camp Mackall for some patrolling practice and I get a sense of what our guys are capable of there.

One morning after breakfast we were moving around the compound when suddenly the loudspeaker system emitted a steady stream of Morse Code. You can immediately identify the communicators in our crew. They all stop short, their faces go blank, and they reach into their pockets for a notebook and pencil.

All the communicators are frantically writing in their notebooks, and the instructors laugh out loud. "See, I told you, and I guarantee not one of them knows what he is writing."

Curious, I slip up next to one of the radio guys and look over his shoulder. The Morse Code is still streaming through the loudspeaker as I read the text in his notebook.

"Slowly, he removed her blouse, the scent of her perfume filling his senses as her full breasts slipped free, shining in the moonlight."

These guys are in their own little world, and it's all dots and dashes to them.

From there, we head back to Ft. Bragg and prepare for Robin Sage. As we completed our planning and prepared for infil, freezing rain began falling and turning into snow. We quickly had over a foot of snow on the ground there at Ft. Bragg. And we were infiltrating into the Uwharrie National Forest in the mountains. The forecast showed no let-up for at least a week.

The decision was passed down after some discussion among the training cadre and a phone conversation with the course commander. Conditions were too severe for forces on Ft. Bragg to train, and sending soldiers into the weather was considered too risky.

But we were Special Forces. There was a strong possibility of injury, but we went anyway. "Do the best you can."

We heard that a lot during the course.

The snow was well over two feet thick in our area once we reached the grid coordinate provided to us. Snow was mixed with freezing rain. When the sun went down, everything froze solid into a single sheet of ice.

We found our "G" force hiding in an abandoned chicken coop with a large fire burning. Step one is to gain their trust and get them to let you into their camp. That was not going to happen that night. We camped in the forest and spent most of the night trying to get through it without freezing to death.

This was before the miracle of Gore-Tex rain gear and special boots, gloves, and lightweight tents. We had rubber raincoats, two ponchos we could use to make an improvised shelter, and whatever leather boots the Army gave us or purchased on our own.

Two trips to Alaska with the Ranger Battalion and two years in the mountains around Bad Tolz helped out here. It wasn't easy, especially since we had several Middle Eastern officers who were really suffering from the cold, but we got through it.

In the morning I was told that one of our two SF instructors hadn't made it and had to be medevac'd for frostbite on both feet. Not a good start.

We worked through the issues to gain trust and get the G-force working with us. We had everything thrown at us, including a small tornado that ripped through the encampment, downing trees, and scattering equipment, but we continued.

The low point came on one mission when we had to hit an objective that required us to cross a swamp to reach it. Not the best time to be in a swamp, but it wasn't over crotch deep. We wouldn't be in it for more than thirty minutes and would be moving the whole time. We could conduct a security halt and get the men dried out once we cleared the swamp. Not good, but it was doable.

But that was our perspective, not shared by our "G" force. Our "G's" were soldiers detailed to do this from COSCOM, the support command on Ft. Bragg. They were good kids, but they were truck drivers, cooks, and clerks, not field soldiers. This was far worse and much harder than anything they had ever been asked to do before. Which was, of course, part of our training in getting them to do it properly.

About halfway through the swamp, one of the G's stopped, threw his rucksack down into the water and started shouting.

"Screw this! I'm not SF, I'm not getting a green beret for doing this, and I'm not doing it anymore! You can court-martial me if you want to, but I'm not going one more step!"

My XO, another brother NCO from the Ranger Battalion and the 509th turned officer led that section of the patrol. He looked over at him and thought for a moment, then walked over and explained the situation to the young man.

"Look, I get it, and you're right, this isn't fair to you. I don't blame you for feeling the way you do. But you have to consider your situation here."

"You are in a swamp. If you don't want to go on, that's fine with me, but you will still be in a swamp and alone with your rucksack filling up with water."

"We are going on and getting out of the swamp and putting on some dry socks. You can stay here if you want or come with us; it's your choice."

After a moment of thought, the G picked up his wet rucksack, and tears rolling down his face, he got back in line and moved out.

It is what it is.

Our class graduated about one hundred and twelve new Special Forces soldiers. We started with almost four hundred and got about fifty recycles from Phase 2 and another twenty from Phase 3 who joined our class as we entered those phases. Not exactly, "One hundred men will test today, but only one will win the green beret," but it's pretty darn close when you consider all the ones who applied but never got past screening. From there, we went into language training. I was headed for the 7th Special Forces Group, so Spanish was on my schedule.

And I would need it.

CHAPTER SEVEN

Colombia

I arrived at the 7th Special Forces Group in the summer of 1989 and was assigned as an ODA commander. Commander was a relative term at that time.

The new SF branch struggled to fill its slots and staff an entire branch. There were new command positions that had never existed before, starting with a Special Force General in command of the U.S. Army Special Forces Command (USASFC). And that meant they needed lots of Special Forces staff officers to fill the headquarters.

New staffs were forming daily, and new Groups were being brought online. 1st and 3rd Special Forces Groups were resurrected. Civil Affairs was formed with slots for Special Forces officers. We needed Special Forces recruiters and logistics personnel.

All this put a lot of pressure on the teams to give up qualified personnel. Especially officers. It was so bad that six months on an A team was considered enough for an officer to be branch qualified as a captain and then you got reassigned. Normally, this meant you would never serve on an A team again.

USASFC was so short of SF officers that they even considered removing the officer slot on A teams and making the Operational Detachment-A all NCOs led by a Warrant Officer.

Now, I've seen plenty of good ODAs that didn't have an officer and did just fine without one. The ODA doesn't need an officer;

SF NCOs are equal to or better than your average company-grade officer in the other branches.

But having SF run from the top by officers who had never deployed as a member of an ODA was not a good idea. Your company, battalion, and group commanders need that experience to function at higher levels. They need an appreciation for what the soldiers on the ground are going through. In my opinion, taking officers off ODAs would have meant the end of Special Forces in the long run.

They created SF Warrant Officers anyway but as the team XO, not the team leader. Training NCOs to be Warrants who would stay on the ODA justified the branch leadership giving the captains assigned as team leader just six months of command time and then be reassigned and moved off the team. Most of them would never serve on an A team again.

The theory was that six months was enough to let the captain understand what the A team did so they could move on to "more important" leadership roles, and the Warrant would provide continuity in leadership. The reality is that six months isn't enough time to go through one cycle of training and deployment, and many team leaders never deployed once with their team before they were moved on to staff jobs.

That was definitely not what those guys signed up for, and it left a lot of disgruntled captains who would leave SF at the first opportunity. But since we were now a branch, they couldn't go back to their old occupational specialty or branch, so they would simply leave the army when their obligation was met.

What made it even worse was that instead of a junior captain or lieutenant being assigned as the team XO where he could gain experience and grow into a team leader, you started as the team leader.

The team sergeants were the most experienced members of the team and the reality was, and I suspect still is, that they provided the necessary leadership continuity and a new Warrant Officer position wasn't needed. I often saw team sergeants leading teams with no officer, and they did just fine without one. I also saw warrant's leading teams with no commissioned officer. Many of them did a good job, but now you had a warrant who was not as experienced as the team sergeant but senior to him, and not quite a commissioned officer either.

Making the team sergeant a warrant if they had to have one would have made more sense to me, but nobody asked me. I always thought that all the warrant program accomplished was cause confusion on the team. We would have been better served to send the guys who wanted to be warrants to OCS and make them commissioned officers. Let them come back as team leaders. That way the good ones could grow into great company, battalion and group commanders.

I arrived at a team that was very skeptical about a new commander, to say the least. I was their third commander in the last year. One of them they never really met. The team was deployed when he arrived, and he opted to go to HALO school rather than join the team in the middle of the mission. Six months passed before the team returned, and he was reassigned as branch-qualified, having never deployed or even met his team members.

The NCOs were not shy about telling me this. There wasn't much point in talking about taking command; I was just along for the ride and was told so flat out. I was advised not to rock the boat and focus on learning how the team operated.

OK, I could see their point, so I decided to ease into this gradually. Starting hard isn't going to work here; better to just wait and see how they do things and where I can contribute.

Our first mission was a training mission to Peru. The team was headed to the Mountain Ranger Camp for training before we went to Peru, so that sounded like a good start on familiar ground. I talked it over with the Team Sergeant and agreed to handle the coordination with the mountain Ranger Camp for our trip.

Our S3 advised me that we were short on training funds for this trip, and I needed to get the camp commander's approval to let us eat in the mess hall without paying the surcharge. OK, no worries.

When we got to the airport in Atlanta and asked for the two vans I had reserved, I was advised that they didn't have any but would give me an upgrade. But all that was available were full-size Lincoln Town Cars, which did not make it easy to convince the camp commander that we were too poor to pay the surcharge when we pulled up into the Mountain Ranger Camp. But we got past that.

While on the rock training, we got word the mission to Peru was off. Terrorists had just blown up a bus full of Russian tourists, the army was in the field chasing them, and they didn't want our help.

We returned to Ft. Bragg, and I went looking for another mission. The Costa Rican police force wanted to build up a new post in the jungle near the border with Nicaragua on the Pacific side, and they wanted SF trainers for the new police force.

That was perfect for us. It was only a ninety-day mission, close enough to trouble to give us a real mission without being neck-deep in it. The team had worked in Costa Rica before near the border with Nicaragua, so it was familiar ground.

I had custody of my son now, which was proving to be a real challenge for a new SF officer. My ex had called me while I was in the qualification course, telling me that if I wanted custody, I could come pick him up, and she would sign the paperwork.

Whether she did that to sabotage my chances of making it through the course or just because she didn't want to take care of our son, I don't know, and I didn't ask. I picked him up and bought a house, but I had a real problem now being a single parent and new to SF.

Fortunately, I had good friends who helped out and family who came in to care for things when I was gone, but I didn't want to be gone too long. Ninety days was long, but the Costa Rica mission wasn't a secret mission where we would be gone without contact for six months to a year.

We made the assessment, met their chain of command, and made our plans. Things looked good. I hired the mother of one of my friends from the Q course to look after my son and the house while I was gone. Everything was going pretty well I thought to myself.

After getting back, I was watching the evening news and saw a news flash about something that had happened yesterday but was just hitting the local news. A candidate for president in Colombia had been running on a platform of anti-corruption, and he had been openly criticizing the drug cartels. He promised that if elected, he would fight the cartels and fire anyone who took money from them.

The news clip showed the candidate, Galan, being assassinated at a public rally. He was gunned down on the platform where he had been giving a speech before a large crowd. President George H. Bush issued a strong condemnation of the assassination and promised all possible aid, short of combat ground troops, to the Colombian government to fight "the war on drugs" in Colombia.

I grinned, thinking someone would be getting a phone call about now.

And my phone started ringing.

The Company Commander told me to come in the next morning and attend a meeting at group HQs. We had a mission, and they needed a team.

All the detachment commanders who were not deployed met in the group conference room to hear the brief. The 1st and 2nd battalions had been passed a mission in Colombia that the 3rd battalion had turned down. The 3rd battalion was located in Panama and was stacked with personnel with the most experience and the best language skills. They fell under the command of the theater CINC and were given the most sensitive missions.

But the 3rd battalion had passed on this one. It called for six SF medics to go to Bogotá to set up and run a combat medic course and establish a self-sustaining medic school for the Colombian Army. An SF ODA had two medics assigned. That would have meant stripping three teams of their medics, leaving those teams without one of the most critical job specialties you need in a combat zone.

Everyone had to be fluent in Spanish since all classes would be in that language, and interpreters would not be available. Many stateside teams didn't maintain their language skills to an adequate level.

We were not at war, but make no mistake, Colombia was a combat zone. There were bombings and killings every day, many of them right in Bogotá city. The training site would be in a Bogotá neighborhood called San Cristóbal, well south of the "red line" set up by the U.S. Embassy. South of the red line was where most of the activity occurred and where the insurgent group M19 had maintained a stronghold for years. M19 had agreed to peace but only recently, and there were other groups, ELP, ELN, and the cartels, of course, who hadn't signed onto any peace agreements. Army units that left the base went out as combat patrols and always made contact in the city.

Not exactly what I had hoped for as a single parent, but the job is the job. I reviewed the requirements, held up my hand, and said, "We can handle it."

"My intel sergeant was a medic before he became an 18F intel sergeant. I've got two fully qualified 18D medics. My team sergeant can set up the instructor training program and establish the new school, and he has cross-trained as a medic. I was a 91B combat medic as a private in 20th SFG, and our communication sergeant can teach the basic first aid classes under the supervision of one of the team medics. We all speak Spanish and have tested out at a 2/2 level or higher. Tactically and technically, this is the best team for the job."

One of the things the guys on my team didn't like about officers was that the missions were handed out in a closed meeting in which the group, battalion, and company commanders received a pitch from the team leaders. A rookie could lose a good mission if he didn't present his case adequately. They trusted the team sergeant a lot more than me to present their case, understandably. But I passed my first test with the team and got us a mission that would usually never come to a U.S.-based ODA.

We made our plan and prepared to deploy. Since President Bush had publicly stated that no U.S. combat troops would be committed to the war in Colombia, we would not be going in as U.S. Special Forces. We would be assigned to the Security Assistance and Training Management Office (SATMO) at Ft. Bragg. This organization used personnel from all army branches to provide non-combat arms training requested by foreign militaries and approved by the U.S. State Department. This was usually logistics, planning, and medical training conducted by unarmed support personnel on foreign military bases where the host nation provided security.

But the area where we would deploy was involved in a shooting war with the drug cartels that was going on right in the capital city. Bogotá had bombs going off every day, the cartels were attacking Colombian army patrols in the jungle and the cities, and no one was entirely sure who in the Colombian military was being paid off by the cartels. There wasn't any area where our support personnel could work where we could be entirely sure they would be safe.

The compromise reached was to use SF soldiers in an unarmed support role. We had the language and technical skills needed to do the job. We also had the tactical training to survive in a fluid security environment working with host nation forces. Our President didn't lie; he wasn't sending an armed SF team to conduct combat operations but sending men who could stay alive in that environment and carry out the support mission.

This would be an urban infiltration, meaning we would go in plain civilian clothes and civilian luggage with no weapons. We would carry sterile uniforms and radios in our luggage, receive weapons and ammunition in country, and carry them concealed for self-defense. Since the U.S. president had openly said no combat troops would be sent to Colombia, we needed to support that and keep a low profile. We would be in uniform on the military base, but off base, we would be in civilian clothes.

We were not on a combat mission; it was a training support mission in a combat area. We were given "imminent danger pay," the new name for combat pay. "Imminent danger" recognized the fact that there were people there who wanted to kill you, and you deserved some additional incentive, without using the politically sensitive term "combat". There was a standing reward offered by the cartels of $25,000 in gold for a dead U.S. Special Forces soldier and $50,000 for a live one.

Secrecy was called for; we would be six gringos a long way from home without help if anything went wrong.

And six guys were all I really had. My senior weapons sergeant was in the NCO advanced course, my engineer was in the basic NCO course, and my other commo sergeant was in another school. I didn't have an XO, and we were short of weapons and an engineer sergeant. Everyone available would deploy.

On the plane ride down, I bought a new book that had just come out to read on the flight. "Clear and Present Danger" by Tom Clancy. Good read, but who would believe that U.S. Army troops would be hunting drug runners in the jungle in Colombia? How does he come up with this stuff?

That book would come back to haunt me later.

We arrived in Bogotá and were taken to the U.S. Embassy for our in-brief by the Regional Security Officer and the MILGROUP commander. The RSO started his briefing by alerting everyone in the room that this was the most dangerous posting outside of Beirut. There were over one hundred people killed inside Bogotá every night, and that was just the normal crime rate. It didn't include the terrorist attacks from the cartels.

The Colombian government was fighting back instead of taking payoffs, and the war was heating up. Simon Bolivar Plaza was the oldest part of the city, the city center, and it marked the start of the "red zone" where the cartels were setting bombs and killing people on the street.

The Colombian Supreme Court building was on the plaza. Insurgents paid by the cartels had seized the building and taken the justices hostage a few years before.

The Colombian Army had used armored vehicles to take back the Supreme Court building, losing the justices in the process. The memory of that lingered, and soldiers were visibly present and heavily armed, with FN rifles, heavy machine guns, and even hand grenades. Simon Bolivar Plaza was the red line where most government defenses were focused.

North of the plaza was where many cartel leaders, military officers, and government officials had homes and lived. Life there was fairly normal, with shopping centers, expensive restaurants, and parks full of flowers.

The top cartel leaders were protesting innocence and blaming the violence on the insurgents and lower-ranking operators. Everyone knew this was BS, but the government wasn't ready to take on top leaders directly. As bad as things were now, that would signal open warfare, and the senior army and police leaders and their families living on the north side were vulnerable.

South of the plaza were the poorer neighborhoods where most of the fighting was happening. Insurgents and cartel operators were attacking police and Army units every night, putting pressure on the Colombian government to back off and let them run their business as before.

We would be based in San Cristóbal, fifteen blocks south of the plaza. San Cristóbal was in the middle of the war zone but also above it on the mountain ridge.

The RSO briefing was for everyone new to Colombia, both military and State Department. He told the room at large that no one was to go south of Simon Bolivar Plaza under any circumstances without specific approval from the Ambassador and the RSO and only in armored vehicles with a military escort.

I raised my hand at that, but he waved me off and said, "I will speak to you guys after the briefing; this is for the group at large."

After the briefing, the RSO gathered us into his office and gave each of us a 9mm Beretta pistol and a fifty-round box of ammunition. We asked for and got three magazines per man, but no holster. Our instructions were to carry the weapon at all times for personal defense and carry it concealed. We were issued Colombian concealed weapons permits and U.S. Embassy ID cards to help us if we ran into Colombian security forces.

"Guys, that briefing was for your information but doesn't apply to you." the RSO said. "We will send you down in an armored vehicle that will drop you off, but I don't have one I can give you to keep. You go where you need to go, anyway you can get there. Please check in once a week and let me know if you are still alive. Good luck."

OK, not very comforting, but it gives us plenty of room to work things out.

San Cristóbal was a charming little barrio of narrow alleys near the top of the mountain range that looms over Bogota' from the east.

The School of Logistics where we would live and run our course was like a castle complete with towers at the corners and built into the side of the mountain near the peak. It used to be a sanitarium for smallpox victims. Charming place.

We were co-located with the intelligence school and right across the ravine from the camp was the M19 insurgent group headquarters.

The M19 had recently signed a peace agreement with the Colombian government, not liking the way things were going and wanting nothing to do with the drug cartels. The ELP, ELN, and FARC were taking money from the cartels and becoming more and more mafia-like organizations, with all the baggage and infighting that comes with that.

But not everyone from M19 wanted peace. Every Friday night they would lob an improvised mortar round over the ravine into the compound, and every Saturday, the Colombian Army would run a combat patrol next door to try and find the guy with the mortar. No one got hurt; it was mainly done on principle, and people kept things civil.

The situation with the cartels was different. They were serious, but they were not in San Cristóbal in force. You would run into them down in the city or out in the countryside. You could usually tell a cartel guy by the nice car and multiple guys with guns and attitude.

ELN and FARC were mainly out in the countryside. If they were in town, they were low profile and dangerous. ELP were college kids; they primarily made the occasional bomb and took over the campus offices from time to time. We would run into them frequently in the cantinas; they loved to talk politics but were harmless to us.

We learned early on to play up the fact that we were medical instructors. In the Colombian culture at that time, medical personnel in the Army were looked down on as not being manly enough to be a real soldier. They didn't have combat medics who went to the field, and they were losing guys who were wounded and didn't make it back to the rear for medical treatment. That's what brought us there. That attitude is what kept us alive. We were not perceived as a threat.

Of course, we didn't mention that we were U.S. Army Special Forces when we were off post. And we didn't mention that we were from the United States. I didn't say American because a Colombian will quickly tell you that he is an American, too, and we in the US don't get to reserve that name for ourselves. It was easier for everyone if we let them believe we were Canadian, German, or anything but U.S. when we were on the streets.

We didn't say we were from another nation. We would just let whoever we met guess and not disagree with them. One man at a local bazaar asked me where I was from, and when I hesitated, he said, "You look German. I know you are not a Norte Americano; they wouldn't be crazy enough to come down here."

"Sprechen Sie Duetsch? Nein, no problem, we can speak English or Spanish, as you prefer."

Most of our time was spent on the compound running our classes. But once a month, we needed to visit the U.S. Embassy and MILGRP to report in, pick up mail, cash a check, or complete other routine tasks.

Movement to and from the compound was challenging but not impossible. The most dangerous area was right in front of our compound gate and at the entrance to the U.S. Embassy downtown.

The Colombian Army kept an eye on things but couldn't watch everything. One of the first times we went out, we saw the

Colombian Army loading two men onto a truck who had their throats cut and their tongues pulled out the slit and hanging down over their throats.

The "Colombian necktie" warned others not to talk to the Army, the police, or the military. The two men had lived in the barrio outside our compound and were informants for the Army. They had attracted the attention of the wrong people.

The barrio streets outside the gate were narrow and ran right up to the gate, with whitewashed walls and red-tiled roofs on one-story buildings lined up like cordwood alongside the road. Every so often, without warning, a narrow alley would separate two of the buildings, creating a maze that you could weave through on your way to or from the gate.

You could take a vehicle, but if someone wanted to take you, they could easily set up an ambush along the winding road leading down from the mountain. They could observe your vehicle leaving and report the make, model, and tag number to others waiting below in the red zone. Anyone who drove out the gate was marked as a moving target the whole time they were out.

Going out on foot was risky, too, but you were harder to spot and a lot harder to track. If you started on foot, caught a bus when you got down on the low ground and a taxi once you reached Simon Bolivar Plaza, you blended in and were much safer. One day I was standing at the bus stop when I saw one of the embassy's armored vans drive by.

It was a civilian full-size van painted to look like a privately owned vehicle, but its radio antennas and dark windows made it obvious.

A Colombian standing near me immediately started ranting to his friend about the "Norte Americanos estupio and their armor vans. Only ten of those in the country; only the Norte Americano drive them, and they think they are safe."

He then described in detail all the vulnerabilities on these vans and how to take one out. His cousin was a mechanic who worked on them for the embassy.

I'll take the bus, thank you very much.

When we first arrived, we were introduced to some of the cantina owners near the base who were friendly but not too friendly, so they were allowed to stay there by the Army and allowed to live by the cartels and the insurgent groups. They were little Switzerland-neutral locations where anyone, cartel, insurgent, or military could stop off, get a cup of coffee, and get an update on the security situation.

We would randomly pick one of these spots, coming or going to see if there was anything to be worried about. They wouldn't tell you who, but as long as you tipped well, they would tell you where and when it would not be a good idea to visit that day.

That was important information. The cartels were applying pressure by conducting random bombings every night throughout the city. They would send their sicarios or hit men throughout the city with boxes filled with paper bags. Each bag contained three or four sticks of dynamite with a detonator and a time fuse attached. They would drive throughout the city and stop near homeless men or others they thought would be cooperative and give them a paper bag with the dynamite and some cash and tell them to set it off wherever they liked.

This resulted in lots of scores getting settled throughout the city and new ones being created. Most had nothing to do with the drug war, but the government had a hundred bombings every night and no way to stop it. But there was one common thread for most of them. No one wanted their friends or relatives to get hurt; people love gossiping.

For instance, the guy with the bomb would stop off and warn people he liked to stay away from the bakery on Calle 13 after ten o'clock because that baker who had pissed him off last week was "going to get what he had coming to him."

And, of course, the people who heard the warning would tell their families. Good tippers were an investment, and the cantina owners and the girls who worked the tables wanted us to come back, so we were included in the friends and family network. Don't ask who would do it; just stay away, and you were good.

During the six months we were down there, I never went a full twenty-four-hour period without at least hearing an explosion or gunfire nearby. Many times we had a front-row seat. But we managed to avoid getting in the middle of it, for the most part.

When you work in an environment like Bogotá in the '80s, you learn to be street-smart quickly, or you don't last. For instance, after dark, only four kinds of people are on the street in the red zone.

The first and most common type were potential victims. They were ordinary people afraid of trouble with their heads down, avoiding eye contact, and moving quickly to get where they were going.

The second type were army patrols or plainclothes police. The army patrols were obvious, and the cops were a little less so, but they still stood out. They were also worried, if not afraid, and tended to stay in groups trying to look in all directions at the same time. They were generally in or near their vehicles and didn't get out unless they had to.

Sicarios, or hit men from the cartels, were the most dangerous type but they were like the army and police patrols. They stayed in or near their cars and usually had some place they were going.

Hopefully, not after you.

That left the common criminals. There are lots of crooks you can run into on the streets. A "ratero" is a pickpocket or sneak thief. Sometimes they would slip an incapacitating drug into your drink or blow it into your face, leaving you stunned and disoriented.

Others would use a "false flag" approach and show you some bogus ID and claim they were police officers, and you needed to come with them. They would take you to some private room or dark alley and shake you down, or worse, introduce you to their friends waiting there. You usually didn't come back from that.

I had one play me like that, coming up to me and waving a large business card with a ribbon and gold seal on it.

"Senor, I am Colombian Secret Police," he said in English. "You must come with me immediately."

"Eso no es una tarjeta de identificación de la policía secreta," I replied.

"Oh," he replied, again in English, "So you speak some Spanish. Very well, my friend, if this is not a Secret Police ID card, tell me what one does like."

"Se ve asi," I replied, flipping out my ID. By this time, we had a Rolodex of IDs from organizations we would work with, so we could show a Colombian military ID to military patrols, a police ID to police, or a U.S. embassy ID for access to the embassy. He went pale and departed quickly. But we clearly needed to adjust our appearance; we were not blending in.

We didn't want to look like victims. There were not enough of us, and we didn't have the guns or cars to pull off being military or police. And they were the ones getting shot at, so that wasn't the best choice. That left criminals. Con men and rateros do not get respect, and they are often victims themselves. Sicario was out; again, we didn't have the numbers, cars, or hardware.

Of all the criminal types roaming the city, "ladrone" best suited our needs. A ladrone is a thief, in Bogotá it's what you would call a mugger. He is usually alone or in groups of two or three. He is the neighborhood bully, usually has a knife or a gun, and is sometimes connected to the cartels in one way or another. You can spot him quickly. He is usually on a street corner or in a doorway, head up and watching, looking for a potential victim. He won't mess with another ladrone unless he thinks he can take him; he will avoid the police and military, but he will come after a victim.

If you are out at night, you choose what type of person you want to look like. We chose ladrone. Black leather jackets are good for hiding pistols and knives. We had credentials for the police and Army if we ran into them, and the sicarios and ladrones would check us out from a distance but usually not approach.

That almost backfired on us in one instance. We had run into one of the Marine guards at the embassy who had invited us to come out to the Embassy Marine house for a drink. We had directions but were unsure of which house to go to when we arrived at the designated place.

I spotted one house with security cameras and barbed wire, clearly more than the other homes nearby. That had to be it, so we walked toward it. Almost immediately, a car pulled up, stopped in the street, and two Colombian men got out on both sides.

This was not normal and a pretty aggressive posture. My teammate and I came on alert, sidestepped one or two paces apart to make us a harder target, and faced the two who had gotten out of the car. But we kept our hands in plain sight and smiled, saying hello in Spanish.

They caught our movement. Civilians or victims approached on the street will tend to come together like a herd of sheep facing a wolf. When threatened, people who get some distance between them are getting ready to fight and know how to.

The man on the passenger side stepped out so his right hand was clear of the door. His partner on the driver's side smiled back and asked me who we were looking for. But he also sidestepped a bit so his gun hand could clear his door through the gap between the door and the hood.

We both spotted their movement and leaned forward slightly onto the balls of our feet, toes pointed toward the men from the car. If you have to draw your gun, you want to be focused on your target and weight forward to absorb the recoil of your weapon.

The movement was subtle, but the man on my left spotted it and moved his hand slightly back toward his hip. You can spot that easily, even if you can't see his hand, by watching a man's elbow. He is reaching for something near his belt line if it moves back or up.

For this reason, I used a shoulder holster. Your hands can be near the center of your body, and your elbows stay down, but you can draw quickly, even while sitting in a car, without indicating what you are about to do. Bogotá had a cool climate, and everyone usually wore a jacket or a suit, so it fit the environment.

I raised my hands to the center of my chest in the "number one" firing position, palms facing the men from the car, and opened in a non-threatening gesture but in a good position to draw. But the man opposite me spotted that and got that "target lock" look on his face.

All of this happened in less than five seconds. In that short span of time, we had rapidly escalated to the point where we were about to be in a firefight solely because everyone present knew what to look for, and we were all reacting to subtle indicators of a potential threat. The guys facing us were clearly well-trained.

I talked the situation down by talking fast and explaining that we were there to meet the Marines, we were with the U.S. Embassy, and we had IDs to prove it. That would either get them to stand down if they were cops or they would shoot if they were sicarios, but at least we would know what we were facing.

It was already down to who could draw first and shoot straight, we needed to know who we were dealing with to do what needed to be done.

Turned out they were Colombian plainclothes police. They had never seen an American approach on foot outside of an armored vehicle, and we looked like ladrones, so they treated us like we were. They allowed me to produce my ID with my left hand, and everyone relaxed. But my right hand stayed in the number one position until I saw that elbow go down.

We had never had plain-clothes policemen approach us like that, but I had been approached like that by armed crooks who were working for the cartels and protecting a street corner. We both reacted to the situation we saw, and it nearly wound up in a gunfight. But we avoided that bullet.

Things were tense in the city, but they were about to worsen. José Gonzalo Rodríguez Gacha, the "enforcer" and head of the Bogotá-based drug cartel had been getting pressured by the Colombian security forces and he was pushing back. Finally, the military and police senior leaders decided enough was enough and they decided to arrest Rodriguez and his men who lived and operated openly in Bogotá.

But Rodriguez had promised to kill the families of anyone who raised their hand against him. The senior police and military leaders decided to move their families out of Bogotá before they made their move. They arranged to send the families as a group on an Avianca flight out of the city to a safe location.

And the cartel placed a bomb on it and blew it out of the sky.

That meant the gloves were off, and the war was on in earnest. We had been working with the police and the Colombian Agrupación de Fuerzas Especiales Antiterroristas Urbanas ("Urban Counter-Terrorism Special Forces Group," AFEUR) based in Bogotá. We had run into both when we were out and about and been introduced through the embassy, visiting their training sites and going through shooting drills with them to stay current.

The AFEUR called us at the compound and asked that we come down. They had a big operation planned in response to the bombing and they wanted medical support.

When we got to their planning area, they explained that the high command was furious about the bombing and that all restrictions on Rodriguez had been removed. They were going after him and all his men that night with everything they had. We were there to provide medical support if needed.

As they went over the target list, they mentioned one target in a barrio near the airport that they were worried about. It was a dangerous neighborhood, and they had no intel on the door, the street, or anything other than the address.

I looked at the map and realized it was one of the barrios I had been to recently. We were comfortable moving around the city by this time, and it was good practice for us to work on city survival skills. During our first month we stayed in the Colombian Army compound, but we needed to come out of the cuartel once a month to cash checks, pick up mail and brief the MILGRP.

And MILGRP had made it plain they were not driving down there to pick us up. So cautiously at first, but with increasing frequency, we went out to get to know the city. It was also necessary to identify Escape and Evasion routes and safe locations in the event everything went to hell, and we had to evacuate the compound.

Which was not impossible. It wasn't long ago that M19 took over the Supreme Court building. Over time, we got to know the city pretty well, certainly better than anyone in the U.S. Embassy. Every so often, I would drop by the embassy and brief the RSO on what was happening in the city, which was greatly appreciated.

After that mission we got more attention from MILGRP, and we started getting other missions in addition to our full-time jobs at the medic course. By this time, we had graduated our first class and were working the best from that class into cadre for the new combat medic school. They were doing well, and we were quickly working ourselves out of the job. One or two more classes, and they could run the course on their own. I was asked to head the security team for a visiting U.S. General and pulled a few other jobs for the command down there, but none of us expected what would come next.

CHAPTER EIGHT

Escobar

O ne morning, the MILGRP commander called and told me he needed two men to fly out of the city that night for a critical mission.

The U.S. Air Force had set up a radar system at a Colombian air base and they were tracking Colombian drug cartel airplanes taking off in the valley. Colombian fighter jocks were intercepting them and forcing them to land or shooting them down. It was not exactly U.S. troops hunting the cartels, but they were in the same valley as Tom Clancy's book. How the heck does he do that?

It worked well, but Pablo Escobar was pissed, and he had called the air base to notify them that if they didn't turn off the radar he would come onto the base and kill everyone associated with it. To reinforce that he was serious, he had tied up a sentry and blew up a transponder at the end of the runway not far from the radar site.

The U.S. Air Force wanted to fly down a security police company to reinforce the site, but the Colombians wouldn't allow it. They said they already had a battalion of Colombian security police there and that should be enough.

The U.S. Air Force had one Security Policeman who was the liaison with the Colombians. He believed the Colombian security force was compromised and some were working with Escobar forces. No one would stand in Escobar's way if he came after them. He was worried and needed help.

The Colombians wouldn't allow a U.S. force to come in, but they would allow two more advisors already in the country. So we got the call.

I talked it over with my team sergeant. He was running the instructor training and he was needed at the compound. Our communicator needed to stay with the team and the primary mission.

Besides, where we were going, they had U.S. communications, and we didn't have anyone to talk to who could come help even if we needed it—at least not fast enough to make any difference. I called back to 7th SFG and let them know what was up, but they were a long way away and wouldn't be able to do much if things went sideways.

We needed one of our medics at the school. In light of where we were going and what we would be getting into, taking a medic on this mission made sense. Our senior medic "JR" immediately volunteered.

That left me and our intel sergeant. The intel sergeant was certainly willing, but the team sergeant had the schoolhouse covered, and this was an important mission. One of the two senior leaders needed to deploy. I would be going on this one.

We wore civilian clothes as usual to travel off base, but this time, we had our day packs as well. Each of us had civilian camping daypacks similar to what you could buy in any sporting goods store, in neutral earth tone colors, brown and green. We also had a nondescript footlocker for larger items that would require a vehicle to get to the airport.

We loaded sterile camouflage uniforms, load-bearing equipment, survival gear, compass, and first aid equipment into the footlocker. This was a special occasion, so the MILGRP sent a vehicle.

We got our briefing at MILGRP, but it wasn't much more than what we already knew. Our mission was simple. Get there, assess the situation, and do whatever we can to protect the U.S. Air Force mission there. There would be no additional support and no additional information other than another phone call from Escobar had come in that morning. He said that if the radar wasn't turned off and the Americans gone within twenty-four hours, he would send his men into the airbase to destroy the radar and kill all the Americans.

We did get one bit of support. The Colombian Air Force general had sent his personal plane to take us to the air base, and it was waiting for us at the airport.

It was a short ride to the airport and a short flight to the airbase in the valley. Now, Bogotá sits 8,660 feet above sea level, and San Cristóbal, where we were based, is almost another thousand feet above that. Bogotá is cool. It's a little cold in the winter when it's raining, and the air is thin. It took a while to acclimate, but at this point, we had been there for over three months, and we could run the mountain from the gate to the castle without too much effort. But the valley was at an elevation of 566 feet above sea level. It was hot, humid, and covered with jungle. We had improved lung power from working out at high altitudes, but working in a jungle would require a different approach.

We landed, met the U.S. Air Force commander, and changed into sterile camouflage uniforms. A sterile uniform is a standard

U.S. uniform but it has no names, unit patches or badges, worn to prevent an enemy who may be in close contact from being able to identify the soldier or his unit. We weren't going to fool anyone here; we might as well get kitted up for the jungle right off the bat. At least if we had to run for the trees, we would be dressed for it.

Normally we would wear civilian clothes off the base where we worked to protect our identity and anonymity. In Bogotá, we didn't draw much attention; there were plenty of German, Canadian, and even Russian people on the streets and more than a few blond-haired, blue-eyed Colombians.

However now we were on a Colombian air base, and everyone knew U.S. Special Forces were coming. Besides, JR and I looked like the Marlboro man with blond hair and blue eyes. There was not much blending in to be done there. But there was no sense in letting everyone know our names. The cartel had a long reach and we both had families in the U.S.

As for weapons, we had our pistols, but that was all. Escobar was bringing more than that to this party. The first question to the U.S. Air Force was, "Do you guys have any spare long guns?"

The Air Force Security Policeman said he could help us out. He took us to a large GP tent in the grassy area next to the runway.

At first I thought this was their command post, but as we walked into the tent, we saw that it was full of crates of weapons and ammunition. Literally full. There were wooden boxes stacked floor to ceiling the length of the tent with narrow walkways between stacks. Case after case of M16 rifles, ammunition, and who knows what else.

"Jesus, how many guys do you have down here?" I asked.

"Well, for security, there is just me. We have four technicians, the commander, of course, and that's about it."

"All this for six guys? You could easily kit out a company with this."

"We didn't know what we would find here, so we brought all we could just in case. We figured that if the Colombians needed more firepower, we could equip them, but we think Escobar is paying some of them off, so we were afraid to pass out guns to them."

Didn't make much sense to me, but lucky us. We started breaking open boxes like kids in a candy store.

We quickly checked out an M16 rifle each, brand new and never fired. Seven magazines each for our Load Bearing Equipment (LBE) and rifles, another fourteen each for our day packs, and we started to feel more comfortable.

The Colombian officer with us pointed to some smaller boxes and told me those contained the "mines."

"We have already dug the holes for mines but were unsure how to set them up. Perhaps you can assist us with these?"

OK, land mines have their uses, but not if you don't have enough personnel to maintain watch over them. An unattended minefield invites trouble. You don't know who might wander into it, and if it's marked, bad guys will take your mines out and set them up somewhere you don't expect to find them. We need to talk about this, but first, let's see what type of land mines we have here.

Questions were answered as we walked up to the boxes. "Claymore mine" was stenciled on the side. Lovely. Yes, we can use those; thank you very much. But you can fill in the holes, you won't be needing them.

A claymore is an explosive device detonated with a wire and a "clacker" that generates a brief pulse of electricity. You can set them up with a tripwire, but they are best employed where you are holding the clacker and waiting for the bad guys to get close. They will shred anything in front of them on a wide arc out to fifty meters, and they are dangerous further out than that.

But the fragmentation all goes forward; as long as you are twenty meters back and lying down, you will be fine when it blows. Perfect for two out-numbered gringos sitting back-to-back.

It's set up on two sets of little steel legs that fold down and are stuck into the ground. You aim it where you want the blast to go, set the detonator, and move back to the rear, reeling out the wire that comes with it. It's filled with C4 explosive and layers of steel BBs embedded in a polymer that disintegrates when the mine explodes, sending all those BBs out in a fan pattern forward toward the bad guys. "Front toward enemy" is molded into the green plastic case just in case you are too dense to figure out which side should be facing out.

If you have detonating cord and non-electric blasting caps, you can "daisy chain" the claymores so more than one goes off when you hit the clacker. In the Q course, we heard tales of one guy taking out an entire company with daisy-chained claymores. No det cord, but we would make do.

As we look over the claymores, I see some large cases stacked on the side of the tent. Opening one, I found a true thing of beauty: an M60 machine gun.

I was a M60 gunner in the Ranger battalion and have always loved that gun. In the right hands, one man can hold off a platoon. And I was a very good M60 gunner when I was in the battalion.

And this was no ordinary M60. It was everything we wanted in a M60 in the Rangers but never got. The barrel was chopped, shortening the gun by a good four inches and shortening the range but making it a lot more maneuverable. Instead of a large boxy fore-end grip, it had a pistol grip for better control. Improved sights. A new buffer system and shoulder stock.

That is definitely going with us. I took two and checked them out, also found LSA lubricant in cans nearby. Two cases of ammunition go into a growing pile.

Finally, we had finished looking over what was to be had in the toy store and felt a lot more comfortable. There may only be two of us, but we were loaded for bear. It was time to look at the ground and do an assessment.

After asking a few questions and looking at the map, we took off with the Security Policeman to see what we were up against. First, he took us on a tour of the base that led to where the Colombian security force had their barracks.

That was the first sign of trouble. The Colombian officers and pilots lived in nice, air-conditioned rooms near the flight strip, the cantina, and the mess hall. All the security troops lived in open-air barracks well out into the jungle up in the hills, with no AC and no cantina. Their food was brought out to them in large pots. Worse, there were no senior leaders there.

The senior men at the soldier's barracks were "Cabo's" or corporals, the lowest-ranking NCOs in their military. These guys were dumped out in the jungle while their chain of command lived the good life on the main air base. Two of the Cabos came out to meet us, plainly not happy to see two new gringos. "Who are these guys?" they asked the SP in Spanish.

"The Special Forces guys I told you about. They are here to help," he replied. There were no smiles at that and no welcome, either. OK, I have a pretty good idea who is working with Escobar, and it's hard to say I can blame them, considering how their chain of command treats them.

Nonetheless, we spent some time with them, asking how they operate, what they do for security patrols, etc. But the Cabos weren't answering questions; they were clearly not happy to see us there.

From there, we go to the radar site. It's set up on the tallest hill in the area, surrounded by thick jungle. Only the radar dish is visible above the canopy of trees, but you can see it for miles. There is nothing hidden here.

There's one road running up the hill to the radar site. The jungle runs right up to a single roll of concertina wire laid out around the vans used to house the technicians and generators, nearly touching the three large generators with thick jungle on the other side of the wire.

Not much in the way of fields of fire. They have built up a sandbag bunker with a single layer of sandbags rising six feet up off the ground on one side of the hilltop, and there is an old-style .30 cal machine gun set up on a stand-up pedestal mount looking over a low wall of sandbags next to the wall of the bunker.

Not sure what they were thinking on that one. The gun can't be fired unless you are standing behind it, but there is a belt of ammo with the machine gun, so it might be useful. But I wouldn't want to be standing up in the open on top of that hill firing it.

On the runway side of the hilltop there is a steep cliff that drops off fifty feet down to a level spot in the jungle below. On the other side there is one spot where we can see the road below where the gun on the pedestal mount is set up to cover, but a thick jungle canopy covers the rest.

OK, it's time to come up with a plan. It's already early afternoon, and Pablo's deadline expires tomorrow morning. He will not likely come during the day; they will more likely come out at night when the Columbian officers are in their rooms by the airfield. So tomorrow night will be show time.

There is one Colombian private on security at the site, so I strike up a conversation with him.

"Do the Cabo's send patrols out at night?" I ask.

"Si, senor, one patrol is always out walking near the fence down there on the other side of the jungle, past the road. They come up here every two hours for water."

The U.S. Air Force had provided plastic canteens, but the Colombian supply sergeant had refused to issue them, afraid the soldiers would lose them. They were all in the supply room; if the troops wanted water, they had to go find it.

OK, the first problem is that we need to get those canteens out and build rapport with the troops. After a trip down to talk to the Colombian commander, we convinced him that it was the smart thing to do. We got him to release some cases of insect repellant that have also been hoarded in the supply room.

In return, we will conduct classes for the officers on the Claymore mines and the M60 machine guns. Pablo no doubt already knows we have them; we might as well let him hear us tuning them up. May convince him to change his mind about visiting.

Next, we needed to clear that hilltop and set it up for defense. At first, the Air Force isn't too happy about that; the jungle is their "camouflage." But after pointing out that they have already received two phone calls from Escobar and that you can see the radar dish from here on the runway, "I don't think the camouflage thing is working for you."

"But the van and generators are well camouflaged from observance from the air," they reply.

OK, you got me there, and it does explain that machine gun up on that pedestal mount pointing at the sky, but I don't think Pablo will be doing bombing runs from the air. It's not really his style. We might want to clear that jungle to see if anyone is creeping up on you in the dark.

We can leave the trees, but the underbrush needs to go. They finally see the logic and give us permission to use the security troops living up in the hills to get it done. My heart goes out to the U.S. Air Force SP, he has had his hands full here.

We go back up and parlay with the Cabo's. We get a few smiles when they see the canteens, water cans, and insect repellant. Their senior sergeant is with us, and orders go out for a work detail to come to the hilltop the next day to clear the jungle.

We still have a few hours of daylight, so it's time to do our recon. The Cabo tells me they don't do security patrols in the jungle during daylight so that we will have the area to ourselves. Perfect.

JR and I go back to the hilltop to do a little Red Teaming on how Pablo's guys might do their approach. That single roll of concertina next to the loud generators running twenty-four hours a day catches my attention.

We step over the concertina and push back the brush on the other side. Sure enough, just past the first layer of the brush is a low tunnel where the underbrush has been carefully clipped away, leaving a well-concealed tunnel through the thickest part of the underbrush right up to the site. Pablo has been busy.

Checking carefully for trip wires or booby traps, we follow the tunnel down the hill toward the road. Extending that line across the road, we re-enter the jungle and move to the fence line.

Beyond the fence and at the bottom of the hill, we can see a white house on a farm, called a "finca" in Spanish, in the distance. There are armed men down there in a mix of camouflage and civilian clothes. Several expensive white Toyota Land Cruisers are parked outside the farmhouse. OK, that's worth investigating further, but not right now. There are at least ten of them down there and two of us up here. We need a little more advantage.

It occurs to me that this is starting to sound an awful lot like the Tom Clancy book I read on the way down here. How the hell does that guy do this? At any rate, we are going to try and rewrite the ending. I didn't like the one in the book.

Night was beginning to fall, so we returned to the vans and introduced ourselves to the techs there. We will stay with them that night, but just in case, we set up two rappel ropes on the cliff in the back. "If you hear gunfire, get on those ropes and down the cliff. You can make your way to the runway from there. Don't wait for us, just move." The drop isn't far; they can go down hand over hand on the rope if needed.

That night passes without incident, but some claymores are set out in key locations just in case. Pablo is true to his word, and we get through the night without incident, but we can expect trouble this coming evening.

The work party arrives and we set them up to work clearing out the underbrush. The carefully prepared tunnel is gone, along with everything that covered it. We leave the tall trees but everything on the ground comes out. It takes most of the day, but we now have three hundred meters field of fire on all sides except for the cliff face.

We then set the detail to work digging fighting positions forward of the bunker. We set up M60 machine gun positions with firing tables and carefully placed aiming stakes that helped focus the gun on the roadway below and the path from the jungle from the fence line. We set up three of these positions, with covered routes into and out of them and the guns set up and ready to fire in two positions.

The last item on the list of things to do is laying claymore mines while the Cabos watch. We take the time to explain to them exactly what a claymore is and what it will do to you. I finish by telling the Cabo's, "Do not come onto this hill tonight. You have canteens and water; there is no need. I expect Escobar to come, and we will kill anyone who tries to come onto this hill after dark."

As we finished, I started a conversation with one of the privates working with us. We talked about the U.S., his hometown, and his family; he is relaxed and happy to finish the day's work. I asked him if he had ever seen Escobar or his men.

He looks around to ensure the Cabo's are not listening, then whispers, "Si, senor, he and his men were at that finca at the bottom of the hill two days ago. They may still be there."

He doesn't want to explain how he knows that and won't say more, but that's enough.

As night falls, everyone clears off the hilltop except for us, our one Colombian security man, and the technicians. The generators are still running, but we have a good field of fire everywhere except for the cliff. I sent JR to watch the cliff face and that side of the road, and I move to cover the generators and the approach from the fence line.

The last thing we do after dark is move the claymores to new positions a little closer so we can see them from our positions—just in case anyone was watching during daylight. Then we waited.

Around three a.m., I see a Colombian security patrol coming down the road. I've seen Colombian security patrols from the barracks before, but these guys look different.

The security patrols I've seen before were not too happy to be out. They either had no cap or their caps set back on their head, carrying their rifles slung or by the barrel on their shoulder. Their formation was better suited to conversation than security. It looked more like a stroll than a patrol.

These guys are different. Every man is in a new uniform and wearing it properly. They are evenly spaced, and every man has his weapon at the ready. They face out and alternate to cover both sides of the road. They are alert, and no one is talking.

Either the Colombian Air Force has dramatically tightened up on discipline and training in the last ten hours, or these guys are someone new. I'm inclined to believe they came from the finca down the hill.

As they approach our hill coming up the road, I slowly rotate the safety bale off the clacker and check the M60 and my M16. Showtime.

I called my partner on our held radio the Air Force provided us. He is alert, but our Colombian security guard has disappeared.

As the patrol reaches the bottom of our hill, I'm getting ready to light them up. Our Colombian security guard pops up from the side of the road below us and stops the patrol. He speaks with them, pointing up the hill toward us. After a few seconds, the patrol turns around and leaves.

And that was that. The rest of the night passed without incident. That morning Pablo called the airbase again and told the

Commander, "Go ahead and run your damn radar; I don't care about the radar or your green berets."

Mention in dispatches by Escobar, I'm honored.

The Embassy and MILGRP are delighted, and the Colombian Air Force no less so. This is the first time in quite a while that anyone has forced Pablo to back down, and the base commander just achieved rock star status with his command.

Our handler at MILGRP called to tell us everyone is very grateful, and we can return to Bogotá. "Great," I replied. "When can we expect the airplane?"

"Well, the general said he needs the airplane, but you guys can make your way back, right? You are green berets."

Lovely. We just made Pablo Escobar back down and publicly humiliated him, and now we need to catch the bus through his backyard.

And that was our only option without a ride. We damn sure wouldn't blend in, but we could catch the bus and let the driver know we were armed and wouldn't be happy if he stopped the bus before we reached Bogotá. Cell phones were a relatively new invention in 1989; they were big and not very portable, and no one on that bus would have one, including us.

It was not the best solution, so I called our handler and reminded him that it wouldn't look good for two U.S. green berets to potentially take a busload of Colombians hostage for the ride into town. However, that's our only option short of walking, and no one on the airfield will drive us.

Our handler took mercy on us and sent a Colombian driver to run the gauntlet with an old beat-up embassy station wagon known as "the hearse" that should blend in pretty well. Armored, but barely, bent up fenders, one headlight out, it looks like something you would see on a country road in Colombia. It's still a long ride through Pablo's backyard, but it's better than hijacking a bus. So we change back into blue jeans and black leather jackets, give up our long guns, tuck away our pistols, and ride past Pablo's crew back to Bogotá with a chicken in a cage in the back for the driver's mother.

We made our way back to Bogotá' without incident and were debriefed at the embassy. Everyone was pretty happy with us and offered to stop the classes over the Christmas holiday period and fly us all back to the U.S. for Christmas leave.

That sounded good to us. We made our way back to the Logistics School to let the guys know we were going home for Christmas. This was very unusual for an SF mission, but the MILGRP commander and the Ambassador had approved it.

On the day we were scheduled to fly out, the cartel set off a major bomb at the police headquarters. Everything was shut down, and no one was flying out anywhere.

We had to delay our flight to the U.S., which connected through Tocumen Airport in Panama City, until the next day, 20 December 1989.

CHAPTER NINE

Panama

Our flight out of Bogotá on December 20 didn't happen. My old alma mater got to Tucumen Airport before I did. At 0103 on the morning of December 20th, the 1st Ranger Battalion made its second combat jump, this time into Panama.

We found that out on our way to the airport. I could hear the radio in the taxicab blaring out something about how it was time to "rise up and help our Panamanian brothers by killing gringos wherever you find them." That was a definite attention-getter, so I had the driver turn it up a bit.

An account of the U.S. invasion of Panama followed that. We weren't going to be flying out that way, but once again, an "official" war was going on, and I wasn't in it.

Not that we didn't have plenty of war to go around right where we were. But in our little war, we were not officially there, or if we were there, we were not involved in combat operations, and if we were involved in combat operations, no one had approved that back in the U.S. No Combat Infantry Badge or decorations for us.

The entire crew was anxious to return to Ft. Bragg and re-deploy to Panama. I went down to MILGRP to talk to our handler about getting back to the U.S. A C141 was coming through Bogotá airport on the way back to the U.S. to pick up a Blackhawk helicopter that had been provided to the Colombians and crashed. We could ride along on that if we wanted to, but "don't get any ideas about jumping ship and heading down to Panama. You are on a mission. Enjoy your Christmas back home, but then we expect to see you back here."

OK, I promise I won't even think about that, at least not until we get to Fort Bragg. Once at Bragg, I called in, and the company commander told me they were headed to Green Ramp to fly down, but we couldn't go with them. Our handler at MILGRP was one step ahead of me; he had called through his chain to ensure we were blocked from the manifest for Panama.

So we headed back to our homes for the holidays. We were not happy, but we were glad to see the family. I was driving down to my mother's house in Alabama, where my son stayed for Christmas.

And the worst snowstorm since our Robin Sage exercise started. Within two hours of starting, Interstate Highway 95 was being closed by the State Patrol, where the snow was just too deep to be safe for driving.

OK, I'm missing another invasion, and now another historic snowstorm is threatening to leave me stranded at Ft. Bragg, where I will miss Christmas with my son. Not happening.

I got off on small side roads worse than the highway, but no one had blocked them off. I went off the road twice, but after a long day of driving I was getting past the worst of the storm. I reached a point where the highway was still open. This was not turning out to be easy, but I got there.

Christmas was great, but soon it was time to go back. I called MILGRP to get instructions on how they wanted us to return. Commercial tickets, perhaps?

No way. The war was still on, and no U.S. military personnel were flying commercially into or out of Colombia. Call this phone number and ask when flight A3409 is leaving. It's a C5 that will be coming to Bogotá, and MILGRP is arranging for you to fly down with them.

Roger that. I made the call and inquired about the flight mentioned. There is a shocked silence on the line, then, "Can you hold, please?"

"Sure," I reply.

Next, I get Officer Friendly on the line, complete with the background echo telling me he has me on speakerphone and is recording the conversation.

"Hi there, how are you today?" he asks in his fake friendly voice. "Who am I speaking to, and where are you, sir? And can you tell me how you know the flight number of that aircraft?"

"OK, look. Sorry if I am stepping on classified flight plans; I was sent this information in the clear, and no one mentioned it was sensitive. My name is Captain Brewer; I'm sitting in my team room at 7th SFG. You are welcome to send someone from Pope Air Force Base down here, or I will go see them; take your pick."

After some more fencing, he was more or less convinced, or maybe it had taken him that long to trace the call. Either way, I was advised that C5 flight schedules are indeed classified, especially to that location, and I could go to the Pope Air Force Base flight operations for more details.

The folks at Pope were not much more helpful, but after a few more phone calls, I finally got them to put me on the line with flight ops at the Air Force base where this flight would take off. After getting a departure time and date on the secure line, I called my guys and gave them the plan. We would meet in Georgia at an Air Force base that didn't get a lot of visitors.

We showed up on flight day in our blue jeans and leather jackets, with shoulder holsters that carried a pistol on one side and a sheath knife on the other. Bogotá was not a tourist-friendly city at that time. That raised a lot of eyebrows, and the Air Force security police gave us a thorough shakedown, along with lots of questions. Where are your pistols, first and foremost?

"Guys, the pistols are waiting for us in Bogotá. We are all U.S. Special Forces, and we are not in uniform because we are going out the back gate of the airport when we get there. Beyond that, you don't need to know."

More raised eyebrows, but we got our ride back.

Our mission in Colombia was drawing to a close. We had our student instructors working, and the guys still wanted to go back to the U.S. and link up with our company in Panama, but the MILGRP wasn't having it. We had to stay for the duration. We would get back to Ft. Bragg again in March.

There was a little tension among the team over this. Six months is a long time to be away from the family; our job was for the most part finished, and life in San Cristóbal on our mountaintop was not exactly a day at the beach. We were missing a lot of creature comforts, like hot water, decent food, and other comforts of home.

One morning, I saw the guys huddled up, holding a very serious and long conversation. Sometimes, this can mean trouble. They were not happy, but they didn't want to talk to me about it, at least not until they reached an agreement among themselves.

I settled back and waited to see what the big discussion was about. Finally, JR came over and announced, "OK, Cap, here's the deal. We all agree, and we need you to help us out." He then produced a single can of Skoal snuff and handed it carefully to me.

"Cap, we are down to this last can of snuff between us. We already called home for resupply, but it will take at least two weeks to get here, so this has to last two weeks for all of us."

"We want you to keep this with you, and don't leave it unattended for a second. Once a day, in the morning, after breakfast, you take that can out, hold it in your hand, and let each one of us take a dip. We only get to use two fingers, one grab, no rooting around, and you don't let anyone hold the can."

"And Cap, just so you don't get a big head, the only reason we are asking you to do this is because we know you don't dip, and we don't trust one another on something like this."

Clearly, we need to go home. Fortunately, it wouldn't be long.

As a going away present, the cartels hit a Colombian government official in the airport coffee shop where we had breakfast minutes after we left to catch our flight. The whole shop was aired out with a MAC 10 machine gun, and there were multiple casualties. Right up until the last minute, Bogotá stayed busy.

Once we were back, we were advised we would be going to Panama, but official hostilities were over. A new Operation was being planned called Promote Liberty. Once again, it would be high risk, classified, and not considered actual combat. The war was officially over, so no imminent danger pay or CIBs, but there were still people there who want to kill you so be careful.

We started gearing up to go down and replace an ODA from another company working in Chiriqui Province. Chiriqui was the westernmost province of Panama on the Pacific side. We would also pick up responsibility for Boca del Toros province over on the Atlantic side, but a senior E7 would be there with three others from another team. While technically they reported to me, they were largely on their own.

I now had an almost entirely new team. Mac and JR went to the Special Warfare Center as instructors. They had been away from home for a long time and needed family time. My commo sergeant went to the 3rd battalion in Panama. Al, my team sergeant, had been selected for the sergeant major academy. That left Goat, my junior medic, and one more man from the team who had been unable to deploy with us to Colombia.

The only remaining man on the team had been in school when we left, and he had signed up for another school when he returned and found us deployed. He would not graduate until we went on this new mission and would miss it, too. We got new graduates from the Q Course to fill out the team and an experienced team sergeant, commo NCO, and Intel Sergeant transferred over to us from another team. We were only given two weeks before we would re-deploy for another six months.

This was especially hard on my son. I can be rightly criticized for leaving him again for another six months, but my brother was now there to be a substitute Dad, and again, I had great friends. I still don't know if I did the right thing, but it's what I did. SF wasn't easy. I either needed to leave SF or do the job I had requested. So I did the job.

Once in Panama, we were in-briefed at Fort Amador, until recently Noriega's headquarters. We would initially be in uniform and openly carry weapons.

Our mission was, again, simple in terms of what we were ordered to do, difficult in execution. In essence, we were a trip wire, set out in remote locations throughout the nation in the cuartels with the new Panamanian National Police to live with them and keep an eye on them. These were the same guys who, before the war, were the Panamanian Defense Force. The PDF looked down on cops and considered them to be men too fat and lazy to be PDF. Telling them to give up their military badges and put on police badges wouldn't be well received.

If it looked like trouble was brewing and they were organizing for a coup, we were to report. Of course, if they killed us before we could report, that would also be a pretty good indicator of trouble brewing.

The war was officially over, and we were supposed to all be friends. A new government was forming in Panama and they needed a chance to hold presidential elections, get organized, and identify who would support the new government and who still wanted to bring back the bad old days of the all-powerful Guardia National.

We moved from Amador to the city of David, the capital of Chiriqui Province, and our headquarters. Officially, I was the head of the Provincial Advisor Team. But we would do very little advising.

My counterpart in the cuartel at David officially supported the new government, but he had been Noriega's personal pilot. His executive officer was suspected of involvement in numerous criminal activities, but there was no hard evidence. The new police officers at this cuartel had been part of Noriega's power base in this side of the country. We had our work cut out for us.

We were not qualified to train or advise police, so we were assigned actual police officers who were also U.S. Army Reserve or National Guard. They would be the new PNP's advisors and trainers. But our primary mission was not to advise and train but to stay alive and report.

We fully expected there would be trouble, and it was only a matter of time until the hard-core Noriegistas left among the new police force made a move to retake the country. We knew they had weapons cached and a secret network where they discussed how to make their move once the U.S. forces were out of the country.

The real mission was to stay ready and head things off by identifying the ring leaders, keep our command in Amador informed of what was going on in the provinces, be ready to step in when things got hot, and, if necessary, call for help. Or escape and evade and stay alive if it was more than we could stop on our own and things got out of control.

With over five hundred Panamanians at our cuartel alone in David, identifying where and when things were getting beyond what we could handle would be a delicate balance. But that was the mission.

That said, there were Panamanians among the force who just wanted a job and never liked Noriega anyway. We needed the U.S. cops to help them learn how a civilian police force operated. With a little luck, we could pluck the troublemakers out of the crowd early on and give the guys who wanted to move on a chance to do so and rebuild their country.

Our U.S. reserve cops were activated for this mission and placed under my command. Which again was a little complicated. I had one Lieutenant Colonel and a Captain senior to me in time in grade, and both worked for me.

To complicate things further, we had three sites we were responsible for and had to occupy full-time, as well as a dozen smaller cuartels we supervised. I posted Goat, another team member, and the National Guard Captain to Rio Serrano in the mountains, about an hour's drive away on the Costa Rican Border. Goat was an E6 Staff Sergeant, but he was in command.

The Lieutenant Colonel stayed with me in David, along with my new team sergeant and two other team members. I sent the senior NCO and two others over the mountains to Boca del Toros.

I could support Goat and his team to an extent from David, but Boca del Toros was over a three-hour drive away, through mountain passes and triple canopy jungle. There was a dirt road that went there, but people who tried to make the trip were routinely stopped and robbed at gunpoint on the road. Our team over there was on their own unless we could get a helicopter from Amador, and that was over an hour's flight away.

As if this wasn't complicated enough, we were also told that we did not have authorization to run any type of intelligence collection effort. Now, that was important to our survival in that we were living with people we knew wanted to kill us, far enough out to be beyond help from anyone in the rear if things got serious.

There was a Military Intelligence Brigade in the canal zone, and its commander was throwing a fit about SF guys walking on his turf. He insisted that he was in charge of intelligence, and we had no legal authority to run intelligence nets.

He was right, even though that was indirectly what we were doing. We were conducting "Force Protection" information collection. He had the mission of tactical intelligence.

The significant difference was that the MI brigade had a few, very few, senior personnel who spoke the language and could move around and survive in this environment. Most of their personnel were analysts and interpreters who wouldn't last one day in this environment. We had dozens of men scattered across the entire country, we spoke the language, and we were in daily contact with the local population and many of the people we were most concerned about.

The brigade didn't have the personnel for fieldwork on this scale. They were organized to analyze the information provided to them and turn that into intelligence for commanders in the field, not go out and collect.

But they had the legal authority to pay people to find things for them. This is called source operations. Basically, it involves finding someone willing to work for you, telling them what you want to know, and paying them or otherwise rewarding them if they bring you what you want. "Tasking" or telling them specifically what you want to know and paying for it is the part that requires legal authorization.

We were used to surviving in hostile environments, and we could fight or run if needed, so we stood a better chance of survival in this environment. However, the MI Brigade wanted to collect exclusively for themselves, and they were willing to put junior enlisted analysts and interpreters at risk by throwing them out into the field unprepared and untrained.

I found this out shortly after we went to work when I received a call from some people I had met in a small town in the mountains called Volcan. Although we were not authorized to run source operations, nothing stopped us from being friendly and talking to folks. That's SF's way of staying alive in the field.

Early on in the mission, we got into the habit of "going out for a walk." We would go out in pairs and walk the streets, talk to the people, and get a sense of what they were worried about. We didn't ask anyone to get information for us or pay anyone to work for us. But there was nothing wrong with having a cup of coffee and listening to whatever local gossip the people we met were willing to tell us.

We made friends, kept a list of contacts, and passed out our phone numbers. If they had problems and needed help, give us a call, and we will see what we can do. We got a lot of calls on things we couldn't help them with, everything from grandmother needing heart surgery and they couldn't afford it, to we had too much rain this year and the crop failed.

But we got to know people throughout our sector, and we were able to help every once in a while. It may be nothing more than helping pull their truck out of a ditch, but it made a friend, and if there were trouble brewing they would let us know, just like in Colombia.

One of my contacts in Volcan called one day and told me there were some very strange gringos up there asking questions, and they were in danger of getting hit by the local Noriega sympathizers or criminals or both.

When Noriega was in power, he set up local self-defense militia groups called "Dignity Battalions" throughout the country and equipped them with weapons. They were paid to keep the local populace in line, conduct beatings and intimidation, and they were suspected of several murders. We knew the local group up near Volcan had been armed, but no one had found the guns.

They were one of the elements we had been warned to watch out for.

They wouldn't mess with us as long as we didn't make ourselves vulnerable and they thought they could take us without anyone else knowing about it. But since we were spread so thin, it was sometimes necessary to go out in two-man teams or even alone. You needed to stay on your toes when you were out but it was still basic street smarts, mainly keep the rest of the team aware of where you were, keep moving, and be unpredictable.

I took the call, checked the team room, and found this was a spread-thin day. I had two men out making the rounds of the smaller cuartels in the province, and they wouldn't be back for a day or two.

We needed to keep the team room where our radios were set up occupied at all times so this meant leaving two men on the cuartel at David. That left one to go out. This was important, so off I went to Volcan.

I got my contact on the phone again and got some descriptions: a short, fat female with red hair and a very tall, black male, well over six feet. They were out on the main street, stopping people passing by and asking them if they knew where the guns were cached.

The female spoke bad Spanish with a "Norte Americano" accent, and the black male said nothing. He smiled a lot and wore expensive Nike Air Jordan sports shoes and a little straw sombrero like they sell to the tourists in the canal zone.

OK, this sounds like an MI Brigade intelligence collection operation. Those two would stick out like sore thumbs in this area and shouldn't be hard to spot. I better get up there and sort things out before someone scarfs them up.

As I drove up the road into Volcan and reached the area leading into the main town, I stopped short near a local cantina. The town of Volcan isn't large. The city government offices sit at the end of the main road, which ends in a T intersection, one side leading off toward Rio Serrano. The road runs uphill to the city offices, with shops and cantinas lining both sides.

From my car seat, I had a bird's eye view of what was unfolding. And what an interesting little drama it was, too.

The two I had been warned about were standing on the sidewalk, trying to stop people passing by to talk to them. You could see people further up the road crossing to the other side to stay away from them.

Three local DigBat leaders we suspected were hiding guns were on the other side of the road. They watched the two obvious Americans from a distance and huddled over their coffee in a serious discussion, glancing over at the Americans from time to time.

A little closer to me were two of the local tough guys. They were known for petty theft and drug sales and had been in and out of the local jail quite a bit. They were in an alley on the Americans' side of the road, one peering around the corner at the two.

A situation was developing here, and I needed to get those two off the street. I started the car, pulled up to them and opened my passenger side door.

"I am Captain Brewer, U.S. Army Special Forces. You are being surveilled by people who intend to harm you. Get in the car, and I'll get you out of the kill zone and explain what's happening.

They both froze, and the female said, "Lo siento pero yo no entiendo Ingles."

"Right," I replied. "Your call. But I'm closing the door in five seconds and leaving you alone." They hesitated a second, and then both got into the car.

I outlined the situation and pointed out the DigBat guys, what I knew about them, and the local muggers. They finally broke out into English, thanking me for stepping in and explaining that they had been sent up there to canvas the local personnel by their chain of command. They asked, and I explained how I knew they were there. I dropped them off at their car and wished them luck. Unfortunately, more and more games of this sort would play out over the next few months.

Our chain of command directed that we change out of uniform and wear civilian clothes to blend in better and avoid an overt military presence. We were not allowed to carry long guns, but we would carry concealed pistols. There was a threat, and we needed some ability for self-defense.

The concealed weapon thing gave me a problem at first. Green berets out and about in civilian clothes with concealed firearms were a shock to the conventional force commanders in the Canal Zone, and some of them were having a fit.

We got a long Rules of Engagement list or ROE and orders to go in plain clothes with concealed weapons. It said to carry your weapon, but carry it unloaded. Carry the loaded magazine in your pocket. If you felt your life was in danger, shout, "Halt, I am armed," and draw your pistol. If they keep coming, take out your magazine, put it in the weapon, and yell "Halt" again. If they still kept coming, jack a round into the chamber and yell, "Halt, or I will fire." If they still kept coming, fire a warning shot.

After all that, assuming you were still alive and your opponent hadn't taken your pistol from you and kicked your ass, you could shoot him, but only to wound, not to kill. With that ROE, our only hope was that the bad guy would laugh at us so hard he wouldn't be able to shoot us.

I sent word back to Amador that if that was the ROE, we would not carry weapons. They replied that we needed to be armed; there was a threat. I explained that with that ROE, all that carrying a weapon would accomplish would be to guarantee that we would be attacked since everyone would know we were carrying expensive Beretta pistols that were not loaded. They could easily take them from us.

They replied that I could direct my team as I saw fit but would be responsible for what happened.

No problem.

You assholes will blame me for whatever happens either way.

I assembled my little crew in the courtyard of the cuartel and gave them their ROE in Spanish in a loud voice. All my guys spoke Spanish, but this exercise wasn't for them; it was for the PNP, who listened closely to what I had to say.

"You will carry your weapon concealed on your person and loaded at all times, magazine in the weapon and round in the chamber. If you are threatened and feel like your life may be in danger, I want you to talk, warn, run away, threaten, and do whatever you have to do to de-escalate the situation and avoid bloodshed. But if nothing works and you are sure you are about to be killed or seriously injured, draw your weapon and kill whoever presents a threat to your safety."

"Let me be very clear here. We are out here alone, and if you pull out that weapon, that will create a situation where you will either have to shoot or back down. If you have to back down after pulling that weapon, whoever made you back down probably won't stop, and if they have a gun, theirs will come out, too. Keep your gun in your holster and get away or talk them down; don't be afraid to run if you have to."

"But if you can't run, and you are cornered, and things are that bad to where you don't have any other choice, draw your weapon and shoot to kill, then get out of there and report back here as soon as possible. If your gun comes out of the holster, there better be a dead body at your feet. Otherwise, keep talking."

I know that's a little extreme, but it was a show for the audience outside. My guys knew how to handle themselves and could be trusted to do the right thing. We had been shot at before, and we knew when to shoot and when to talk. The show was for the Panamanian PNP who might think about pushing my guys into a corner. I wanted them to know it would end badly for them if they went there.

It paid off about a month later. My guys in Boca del Toros had done the same thing and ensured the ROE was well known to the police in their compound. Their compound, like ours, housed the local jail as well. Panamanian jails tend to be square buildings with an open square in the center. The cells are in a line around the open center under the roof.

Conditions were not good. The food was bad, and there was no air conditioning. Panama is hot and humid, and men locked up in these conditions are, to say the least, not happy about it. We were pressing for better conditions, but that took funding. Things were getting better, but not fast enough.

On one particularly hot day, the prisoners had enough and began to riot. They were usually not locked in their cells during the day, and they were trying to batter the big wooden doors leading into the courtyard open with a makeshift battering ram.

Over a hundred men were shouting in the stifling heat, steam rising off bodies as they threw rocks at anyone who appeared on the wall while they worked on battering down the door. My guys were on the other side of the door, trying to talk them down. They explained that this wouldn't fix anything and if they didn't stop, someone was going to get hurt. But the prisoners were beyond reason, and they wanted blood.

The Panamanian guards lined up behind my guys at the door and loaded their rifles. We knew full well the guards would like to shoot us as much as the prisoners, and they had proved that in David when the team before us had a similar riot. The guards began firing directly at the crowd of prisoners with U.S. Special Forces among them, trying to settle things down.

The guys in David had resolved that one by opening a case of CS grenades and throwing every CS canister they could get their hands on into the compound. Prisoners and guards alike gave up fast when that happened. But the lawyers got involved and wanted to put our guys in jail.

The police are allowed to use CS, but for the U.S. Army to use CS, we needed National Command Authority permission from the president. The lawyers had decided that using CS constituted the first use of chemical weapons by a military unit.

Needless to say, we didn't have any CS this time around. As the doors finally burst open and a hundred rioting prisoners burst out, the guards behind my guys let their bolts go forward and aimed toward the crowd with my guys right in the middle. Faced with this dilemma, they did the only thing they could.

They drew their pistols.

And I swear to God, one hundred rioting prisoners dropped their clubs and rocks and threw their hands into the air. Stunned at this reaction, my guys turned to the guards to tell them to hold their fire, and the guards threw down their weapons.

Later, I asked the Lieutenant in charge of the guards that day why they had dropped their weapons. My guys had not pointed their pistols at them; they simply looked toward the guards and told them not to shoot. The Lieutenant replied, "Capitan Brewer, we knew your ROE, and when the gringos drew their pistols, we knew someone was going to die. And it wasn't going to be us."

Alrighty then. That worked out pretty well.

In some respects, the civilian clothes order made things easier for us. We were initially operating in uniform, wearing our green berets and carrying rifles and pistols openly. The people loved us and would come up to us readily to talk. But it was always a problem back in the Canal Zone.

Coming into Amador, we would be stopped by the MPs every time. They wanted to search our vehicle, our pockets, check our authorization to carry weapons, and even check to see if our rental cars were properly dispatched by the motor pool.

Since the dispatch was only good for thirty days, we had to return to Amador every thirty days to re-dispatch. We could be up to our ears in armed bad guys, in a hurricane, with the building on fire, and no one would come to help us. But God help you if you didn't return to Amador in time to renew your dispatch.

It got so bad that the MPs stopped one of our guys and arrested him on the spot for "transporting two prostitutes in a government-dispatched rental vehicle." Of course, when the female Army chaplain and her assistant produced their ID, they let him go, but it was getting silly.

One incident summed up just how ridiculous things got, even more than the one with the chaplain. One of the other ODAs had been at a police checkpoint when a truck ran through the roadblock and tried to escape.

We had a lot of restrictions on what we could do. The bottom line was to "observe, monitor, and report" and stay out of the middle of things. We were not to touch anything that involved drugs. The DEA had staked that out as their exclusive turf, and we were ordered that even if we saw drugs being sold, not to intervene, just walk away and report it.

But it's hard to tell a group of young hard chargers to look away when they see the action start. The PNP pursued the truck, and the guys from the ODA followed. It turned into a high-speed chase, with PNP firing out the window at the truck. Finally, the truck overturned and went off the road, and the PNP vehicle slammed on its brakes. Our gallant SFers close behind wound up rear-ending the PNP vehicle.

The driver and the passenger in the truck came out with guns blazing, and the PNP, as well as our SF guys, returned fire. The gunfight ended quickly with one wounded but no one killed, but the reason for them running the roadblock was readily apparent. Large packages spilled out of the back of the truck full of cocaine.

Now, you would think this would earn the SF guys an atta boy for stopping a major drug shipment. But no. The DEA had been reporting that the drug trade was under control following the invasion, and no major shipments were moving through Panama. The truckload moving through an area they never visited made them look foolish, and they were raising hell about U.S. Army SF getting involved in drug investigations.

The briefing at the inquest, during which our company XO presented a defense, pretty much summed up the political BS that had reached epic heights on this one.

"And in closing, sir, while it is true that our men were involved in a drug-related investigation, they did not know drugs were involved at the time; it only later became apparent. Their use of firearms was in self-defense of both themselves and the PNP officers in that the suspects fired first. And while it is true they exceeded the speed limit in a government contracted rental vehicle and were involved in an accident with a Panamanian government-owned vehicle, on the plus side, their dispatch was up to date, and they were wearing their seatbelts."

Amazing. They let them go with a stern warning, but only because they had a current dispatch. You have to wonder about priorities with these people.

So, we went to civilian clothes to lower our signature. And it worked. We didn't draw as many reporters in the field, and the MPs left us alone. I came back through the same gate in Amador where we had been hassled for an hour over our vehicle dispatch and weapons authorization, this time in civilian clothes and carrying a truckload of AK47s we had recovered from the field under a tarp in the back of the truck. All we got was a wave through.

It's not that we were in hostile territory; it certainly wasn't as dangerous as Colombia. The vast majority of the local people were happy to see Noriega gone, and the power of the PDF was cut back. They were still afraid the PDF would try to take back the country, and they welcomed our presence.

Most of what we knew about what was going on didn't come from paid intelligence agents, it came from local people who stopped us on the side of the road to talk to us. We explained that when we were accused of running intelligence networks, and apparently, the MI Brigade was trying to do the same thing, albeit pretty clumsily.

We would continue to run into the MI people occasionally, but now they focused on us. We were viewed as a threat to "their operation." They were tasked with following us and getting the goods on us.

Not that we were hard to follow. We stuck out too, and we made no effort to hide. Our technique was to go out and be seen and see who would come up to us to talk. We didn't push, we didn't press, and we didn't ask too many questions. We just welcomed whoever came up and let them talk about what they wanted to talk about.

Of course, there were people out there who did want to hurt us, many of them living in the cuartel with us. But they wouldn't take a chance of attacking us on the cuartel or anywhere else where they could be identified and held accountable. Inside the cuartel you were pretty safe unless the coup started and everyone came out to fight.

Outside, it was a complicated game of reverse hide-and-seek. You had to keep an eye out for the inevitable tail, look ahead for an ambush, stay visible to as many witnesses as possible, move rapidly through lonely areas, and watch your back while you did.

Spotting the MI types was simple, most of them were white guys with military haircuts and a scowl on their faces. The Noriegistas were a little harder to spot, but the people knew who they were and were afraid of them. They would part like the Red Sea in front of Moses when Noriegistas appeared.

One night, I was out on a walk and found ten expensive Toyota four runners parked outside the house of a local guy we knew was not friendly. Most of them had license plate numbers that showed they were from the capital city. Standing nearby was a big gnarly looking guy with long hair, a scraggly beard, camouflage jacket, blue jeans, and a big sheepish grin. He had his right hand behind his back.

OK, this bears closer inspection. I walked up and said hello to big ugly, who, luckily, was on the other side of a steel bar fence. He walked over, still grinning and nodding.

I introduced myself and held out my right hand to shake hands. I usually carried my pistol with a right-hand draw, but I had my Tanto in a sheath on my left.

Still grinning like a fool, he froze momentarily, put his left hand behind his back, and brought his right around to shake.

Making sure he reached through the fence to my side, I grasped his right hand and said, "What do you have behind your back in your hand, my friend?"

Still grinning, he brought out a cut-down machete that had been sharpened into a nine-inch dagger and put it down on the hood of the car nearby.

OK, now that we are past that, I asked him what he was doing out here. He said he was guarding the cars for the important men inside. "OK," I said, "Who are these important men?"

"Oh, I don't know their names, senor. I just protect the cars."

"Do you live around here?" I asked.

"Oh, no, senor, I am from Panama City. I rode up with them."

"OK, you don't know their names, but you rode four hours with them from Panama City. No problem. I'll just write down the tag numbers, and we will check who is here."

"Oh, no, you must not do that, senor."

"Why, my friend. We are all friends now, no?"

Big Ugly didn't have an answer for that. I copied down tag numbers, smiled sweetly at him, and went on my way. I ensured I kept the windows of the house and him in my field of vision until I was out of sight in the darkness.

Those tag numbers would open another chapter in this story.

CHAPTER TEN

Crisis

We sent the tag numbers back to our B team in Amador. Other teams saw similar things in their sectors. Expensive cars were showing up at the homes of known troublemakers late at night for secret meetings.

We connected with the old police force in the capital and SF guys working with the new PNP. The PNP guys in the Panama City were working to build a new country and cooperating with us. Most of the troublemakers were out in the countryside, where we lived and worked. The PNP ran the tag numbers for us and gave us what they knew about the owners.

Our B team ops sergeant had deployed to Panama with his own Apple computer. All of us had been issued new laptop computers, the first time we had seen one. We didn't know what they were for at the time, but we were told to write all our reports on the laptop and send them electronically to the rear.

This was all new to us, and most of us didn't like the laptops. They were slow and used something called Office Writer. It had a C prompt where you had to type in commands to get it to do anything. You could type in four commands and go get a cup of coffee, and it would probably be done by the time you return. Maybe.

Before, we wrote reports by hand and encoded them to transmit by radio or sent handwritten reports by courier. If you were very high-speed, you might have a manual typewriter. But that was about it.

But orders were orders, and we got used to the new laptops. And the tag numbers showed us why it was important.

Our B team ops sergeant set up a searchable database and started logging all the sightings by tag number with a date, time, and location. We quickly started to get a picture of who was talking to whom. We could send in a tag number and get back an email identifying the owner of the vehicle with a complete list of known associates, the last reported sighting, and who he was likely to be talking to where we were based on past reports.

We had another new device that came in handy for receiving and sending reports in the field. About the size of a pack of cigarettes, it fit easily inside a jacket pocket. It had a data connection to plug it into the radio or the laptop with an adapter. It could encrypt the computer email and send it out safe from intercept.

It also had a microphone and speaker on the back. It could send or receive an audible burst transmission that would then be decrypted and displayed on an LED panel as script. You could also type in a message on a small keyboard on the front under the LED panel.

We discovered that by putting some moleskin adhesive bandage around the microphone and speaker to get a good seal, we could send and receive encrypted messages from a pay phone booth. This was 1990, and laptop computers and cell phones existed, but barely, and they were rare. Tactical radios were big and heavy and weighed a good twenty pounds.

Combining these two new pieces of technology gave us options we had never had before. We could move about in urban areas, communicate securely, and have access to information at our fingertips. It opened a world of possibilities.

And just in time. The old-guard PDF was getting restless. The elections had been held, and Endara became the new president. Every day that went by the new government was getting things better organized, and the opportunity to seize power by force was slipping away.

The U.S. government had made a conscious decision to keep the PDF in their barracks with their chain of command intact unless there was evidence of criminal activity. This had proven to be a smart decision and kept a lot of young men who were trained as soldiers employed and working toward building a new Panama. Disbanding them would have left them with no jobs and no options and almost guaranteed an insurgency.

But some in the ranks still wanted to bring back the bad old days, and they were getting ready to make their move.

We started to see some attitude among some of the police in David. There was nothing you could point to as open resistance, but cooperation was getting harder to come by. The people in the town were getting nervous as well. The commander of the cuartel started locking his door at all times, and we couldn't get in to see him unannounced.

The rank and file started openly carrying their rifles around the compound. Normally, rifles were not issued or needed. And we were starting to get reports of bombs going off. Not in town, and no one was hurt; nothing significant was damaged. They were blowing up telephone poles leading into the city, first from fifty miles out, then a few miles closer, and finally, one went off in the bottom of an old abandoned swimming pool in the city.

They were using sticks of dynamite, not military explosives. And it was pretty clear they didn't want to hurt anyone. But the people were getting rattled, and it was being done to show the gringos couldn't do anything about it.

The police investigators would shrug their shoulders and say it was a mystery. It was clear they knew more than was going on, but they were not talking.

About this time, the local prosecuting attorney asked us to come over and talk to him. He was working with the government and had been supportive. But he was concerned, and he had been threatened by men who called him and told him to drop his current investigation, or he would be killed.

We got to his office, and he told us he was about to arrest a local businessman who retired from the PDF. The prosecuting attorney had evidence that this man was involved in the murder of Hugo Spadafora. Spadafora had opposed Noriega and had been caught, raped, tortured, and murdered in Chiriqui Province. He had been decapitated, and his head was never found.

We went with the attorney on the arrest and accompanied the prisoner to the David Cuartel jail. He was met by one of the senior police officers, who shook his hand and told him, "Do not worry, my brother; everything will be OK." It was clear where the new cops' sympathies lay.

The next morning, we found an envelope with color photos of Spadafora's headless, mutilated body on our doorstep. Pretty clear message.

Tension continued to mount. There were more and more locked doors. Our phone line went out every day. We set up our satellite radio, but the antenna would get knocked off the roof as soon as we looked away.

This all culminated with our boss in Amador calling me one night and giving me the code word to initiate Escape and Evasion procedures. But he followed that with, "Not just yet. But be ready." The coup was about to begin.

We had one MI type with us by that time, an attractive young woman who was a translator/interrogator, spoke great Spanish, and had been sent to live and work directly with us as a compromise of sorts with the MI Brigade. She didn't have any field training or training in setting up a source network, but she was willing to learn and had the authority to run sources and pay for intel. She became an invaluable team member and contributed a lot to the effort. But we would continue to have trouble with the MI Brigade.

I got a call from my boss in the rear, telling me the MI Brigade commander was coming out to talk to us. "No problem, we can brief him on what we are doing. We know the radio room is bugged, so we may need to take a walk or a drive to get into the sensitive areas."

"No, you don't tell him anything about our operations, period. Give him the unclassified visitor brief in the team room, but that's it."

"You know he isn't going to like that. He's likely to leave here a very unhappy camper. Besides, we have his agent here with us, and she knows everything we are up to anyway; there isn't anything we are doing that she hasn't already reported."

"It doesn't matter what she reports; you don't report to him. We don't want to set a precedent here. He is trying to take control of the operation, and we can't have that."

OK, I've seen firsthand what this guy is doing with his people in the field, and granted, I don't want any part of that either. This isn't going to be pretty, but we will see how it goes.

That morning was particularly touchy. Once again, one of the PNP knocked our SAT radio antenna off the roof. We go back up and set it up again two or three times, but if we leave it unattended, it comes right down again. If we complained, they smiled and told us, "Must be the wind." But there is no wind.

One of my guys comes in and tells me he just saw three PNP carrying AK47s into their barracks. They are not allowed AKs; they have the Chinese version of our M16 that they hate, but it's what is authorized. When we question them, they deny everything and refuse to allow us to check the barracks. The PNP are becoming openly belligerent and arrogant. Its pretty clear things are building to a head.

And then the MI Brigade commander shows up. What a perfect day. But I take him into the briefing room and give him the one-over-the-world brief. As expected, he isn't happy.

"I don't need to hear that BS; tell me about the guns."

OK, we know little about guns other than what I saw this morning, and I've been instructed not to tell him that. I advised him that this room cannot be considered secure, but if he comes with me, I'll tell him what I can.

We move out of the briefing room and to the room I use for my quarters. We checked it thoroughly when we moved in, and it's isolated from the rest of the building. There are no common walls with the PNP area like our commo room next to the PNP commanders' office. The PNP gave us the commo room they did for a reason, and we were certain they were listening to every conversation that went on there. We used code words and encrypted everything that went out of there.

But my room was probably as safe as anywhere on the compound, and normally, we never held meetings there. It had been the PDF interrogation room, with a chair bolted to the floor and a hand crank generator next to it when we moved in. The locals were terrified of the room. I slept there to let them and the PNP know that the room would not be used for that purpose again. But it was quiet and had the strongest door and walls of any room we had.

Once inside, I sat down and explained the situation to the Colonel. "Look, sir, I don't have the authority to discuss our operations with you. I apologize for that, but those are my orders. Whatever is going on between you and our chain of command, I don't want to get in the middle of it, but the bottom line is we don't hide anything from your agent, and she can brief you fully on everything we know."

"That's BS. You knew I was coming out here and should have called for authority to brief me. I don't want to hear it from her; I want you to brief me fully right now!"

"Sorry, sir, but I called my higher, and they said not to brief you. It needs to come through your agent, not me."

"Captain, I am giving you a direct order to brief me right now! I want to know what you know about guns and anything else and don't give me any more of this crap. That's an order!"

"Sir, with all due respect, that just isn't going to happen. But I will tell you we have a situation here, and you and the team are at risk. I have observed things this morning that concern me deeply, and we know we are being surveilled. I cannot guarantee the security of this room or this compound. Under the circumstances, and in light of your attempt to use your rank to force me to disobey my orders, you are adding to that risk to my team. I'm going to have to ask you to leave now."

"And if I refuse?"

Deep breath. OK, he won't budge, and I don't have time to play, "I've got an eagle on my shoulders, and you don't."

"Sir, I'm sorry to hear that, but if you refuse to leave and continue to place my command and yourself at risk by pushing us to openly discuss sensitive information in a non-secure area, I will have to remove you by force."

At this point, he folds his arms and says, "And just how do you intend to do that, Captain?"

"Well, sir, off the top of my head, I'll have the two E7s standing behind you tie you up, gag you, we will put you in the trunk of your vehicle, and drive it and one of ours back to Amador with an armed detail to protect you. I hate to do that, but we can't have the PNP see us carry you out kicking and screaming. But one way or another, we are done talking now, and you are leaving."

He gives me a long, hard stare, and I return it until he blinks. He gets up and walks out. Outside, the young captain who came out with him is waiting.

"Sorry we couldn't be of more help, sir. I hope you understand."

"I understand just fine, Captain. You have not heard the last of this."

I step forward, smile, and hold out my hand to shake. "That's your choice, sir. But goodbye." My two guys step forward with me, big grins and waiting for instructions.

The good Colonel wants to say more but reconsiders, gets in his vehicle, and slams the door. And he's right; I haven't heard the last of it. As soon as he returns, he tries to file court-martial charges against me.

First, he starts with disobedience to a direct order. But my chain backs me up. I was following orders from a Colonel I worked for, and he was out of line.

He follows with charges of assault. I threatened to tie him up and carry him off by force.

But I didn't do it; I simply pointed out that we were in a tactically dangerous situation, and his actions endangered my command. I politely asked him to stop and leave but pointed out that I would have to take action to protect him and my command if he didn't. That charge doesn't fly, either.

Finally, he charged me with sexual harassment. His orders to his female agent were to go out alone and "gather intelligence." She's a great asset to the team, but she isn't a trained intelligence operator. She is an interpreter. When she arrived, she didn't even have ammunition for the pistol they issued her, and she had never fired it.

I explained the situation to her and pointed out that we couldn't hold her there in the team room, but if she went out alone, she could expect to be taken somewhere quiet and never be seen again and we wouldn't be able to stop it. I briefed her on Spadafora and what happened to him.

We didn't want to stop her mission; on the contrary, we wanted to help. But it would have been better if she went out with one of us until she knew the area and learned some street smarts, beginning with how to use that pistol. After some time, she became as sharp as any of the guys and a team member. She did go out alone eventually, but not until she had time to learn how to do it and come back alive.

Those charges also fail to fly, but only because she backs me up and refuses to go with the story the Colonel gives the investigators. But her career is over. Eventually, he will relieve her and pull her out of the field, although she will stay with us for another month. Damn shame, too, she was the best operator he had, and she was sending in the best intel he got. It was all about power and who was in charge.

Our tactical situation resolved itself. Whatever they had been planning got headed off above our level, but it wasn't over.

The attempts to draw us out where we would be vulnerable were getting more direct and overt. We were approached and told by a walk-in that he knew where the guns were, but we needed to come with him at night. And they wanted me and our female agent to come, but no one else.

We did go to the meeting as agreed, with several of my guys quietly positioned out of sight with M16s to overwatch the meeting. The guys who met us were grim-faced and wanted us to get into a car and go with them alone to an undisclosed location, and there were four of them. Sorry, guys, it's not happening, but thanks for coming out. We now have a few more names and leads. These guys are all Dignity Battalion hardcore.

The crisis continues to build, and it's becoming more obvious that we are reaching a breaking point. One night, our female MI agent comes into the team room in tears. She has gotten pretty good at moving around independently and travels alone quite a bit now. She has good survival instincts and can spot a trap; she has developed into a real operator.

But she didn't anticipate what was waiting for her at the gate that night. The two PNP gate guards held her at the gate for the last half hour, searching her vehicle, then her body, groping her, and making very specific comments on what they would soon be doing to her. They have never dared to be this bold before.

Our SAT antenna is knocked over again, and the telephone line is cut.

I've had enough. I tell our commo sergeant and weapons sergeant to take the SAT radio out into the middle of the compound and send a message to the rear about what's happening. And I tell the rest of the team to draw their M16s out of the rack and take up positions to protect the radio. Shoot anyone who tries to stop you from sending the message.

I head to the gate. I've got a few words for the knuckleheads who molested one of my team. They are grinning at me as I walk up. They won't be grinning long. I stop and glare at them for a moment. More PNP are spilling out of the barracks behind me to watch the show. They don't approach the radio, but they are gathering behind me. Good, I need that message to get out. I ignore them, thinking about what I will say. Anger won't quite do it; I need to get control and put them off balance. These guys think they have the upper hand, so I need to counter that.

"Why did you stop and search a member of my team so rudely?"

"We search everyone who comes in here gringo, you are no different. You have no power here."

"Really? Let me be clear: you put your hands on a member of my team again, and you will deal with me. And you won't be smiling when I am done, I promise you that."

At this point, a member of the crowd behind me shouts out, "Hey, gringo, you are a long way from the Canal Zone, you better think about that."

"Really? What month is this?" They were not expecting that, and they looked confused.

"What do you mean?"

"What month is this? November, is it not? And what month comes after November?"

"December, gringo. What's wrong with you? Are you crazy?"

"December, yes. And what happens in December?" Now everyone gets very quiet. December is the anniversary of the invasion. The reference is obvious.

"December, you know, that's the Christmas season. And what happens during this season." Everyone relaxes just a bit, but their confusion is still evident.

"December. That's when the big guy with the red hat comes down here bringing presents for everyone. Sometimes it's Santa Claus. And sometimes it's the 82nd Airborne Division. YOU had better think about THAT, my friend. Now, get back in your barracks before someone gets hurt. Because if one of us gets hurt, I promise none of you will escape the consequences."

It works, and everyone returns to their barracks, but the cuartel commander is already on the phone dropping dimes on me. "Capitan Brewer is still trying to fight the war; he doesn't want peace."

That opens a major investigation. It's obvious what was happening, but there isn't hard evidence to support either story, as usual. So, there is a compromise.

The commander of the David cuartel is relieved, and a firm supporter of the new government comes out to replace him. The XO we have always suspected of drug use and murder is arrested. Things calm down a bit, but it won't last.

In the rear, our company commander is moved to another assignment in the middle of the mission. It's unusual, but we have a new company commander with a lot of experience in Panama. I am tapped to return to the Canal Zone and be his XO to help him transition and as part of the compromise with the Panamanians. I'm losing my ODA, but I will be at the center of what's been going on back in the rear and get to see the bigger picture.

And just in time for the big finale.

Chapter Eleven

Ft Amador

No one wants to leave an ODA if they genuinely love SF. That's where the work is done and what SF is truly all about. I was certainly not happy about it. But I had been incredibly lucky, too.

We were still short of SF officers, and a normal ODA command tour took only six months. Many SF officers never deployed on a training mission with their teams, let alone an operation. I had been a team leader for sixteen months and was the senior team leader in the 7th Special Forces Group. I had not one but two six-month deployments into dangerous areas and one named Operation under my belt.

The new Special Forces Command wanted me back at Ft. Bragg for a staff tour. However, the senior leadership on the mission in Panama moved me to the company XO slot and kept me in the field to help the new company commander. That did not make USASFC happy, and they tried throwing their weight around to force me to be transferred back immediately.

But we were deployed in theater, and the theater CINC informed USASFC that they had no authority in his sandbox. This would not get me a warm reception at my new job when I finally arrived, but I was still in the game for now.

We had seen more activity in our province than any other team, and in many ways, Chiriqui was the center of gravity of the resistance. What I had seen there would be invaluable to the

170 | Ft Amador

command back at Amador while we tried to head off problems. But I would also get to see what was happening in the rest of the country.

We had six ODAs spread out over fourteen sites in the country, from the Costa Rican border in the west to the Colombian border in the east. Many people in the U.S. don't realize it, but the country runs west to east, not north to south. The isthmus curves to an east-west direction right about there, and the canal runs north to south, not east to west.

The country is roughly four hundred miles wide, border to border, but the last fifty miles to Colombia in the east have no road. The area is a swampy jungle that not even the natives who live there like to enter. We had two sites down there, one on the Pacific side and one on the Atlantic.

As part of my new duties, I would need to carry cash out to each site every month. Every ODA had an operating fund to buy gas, food, and construction materials to repair or fortify the safe house, whatever the team needed. They are out there on their own; if they can't get it there, it has to be flown into them, or they have to do without.

Once a month, I would go down to the Army finance office and draw a little more than a quarter of a million dollars. Each of the fourteen sites would get an operating fund of twenty thousand dollars. This would pay for fuel, food, rent, and whatever the teams needed to improve their site or conduct the mission. Everything had to be supported by receipts and examined monthly to ensure the money was spent appropriately.

But the delivery of the money was a challenge. We were spread very thin, so I would have to travel alone. Again, the countryside was not especially dangerous, and the locals were friendly. But there was a criminal element, as there is in any country, who would be very interested in that much money, and as always, there were the Noriegistas.

Travel security was paramount. Going out with an armed convoy and overt security would guarantee that someone would try to attack the courier sooner or later. Overt wasn't our style. We traveled alone more often than not to and from the Canal Zone. We were careful, we knew the language and the culture, and we were used to working in places much more dangerous than this.

I would deliver the cash to most of the teams by car, driving out and making deliveries along the route to David. ODA commanders would take the cash for their sites and manage it from there.

I learned early on not to stop and to keep going from one site to the next and rest or eat there. On my first trip out, I stopped to get something to eat at a cantina. You don't leave a quarter of a million dollars in an unattended car, so I carried the small day pack I used to transport the cash. I was armed and used to moving around alone, and I didn't plan on staying long anyway. Grab some food and go.

But that much cash in brand-new bills gives off a distinctive odor. Especially in a hot tropical climate when you bring the bag out of an air-conditioned car into the heat. Sitting at a table, I started to catch the distinctive smell of new bills I remembered from the finance office where I had drawn the cash. And I could see other patrons lifting their heads and sniffing with a puzzled look as they tried to place that smell.

I moved out smartly before they figured it out. No more rest stops.

Deliveries to the eastern side of the country were a different matter. The east side of the country was considered an "impenetrable jungle." That's a relative term, of course. Having been through the Jungle Warfare Course three times and spent more than my share of time in Panama's jungles, I knew a little about the subject. But I must admit, you didn't want to try to move through this jungle. There were roads and towns along the Caribbean side of the country in the west near Costa Rica, but in the east near Colombia, the isthmus narrowed.

The terrain along the coast in that part of the country was slightly above sea level, and much of it wasn't above sea level when the tide was in. Most of it was a thick mangrove swamp and triple canopy jungle. A mangrove swamp is a thick tangle of mangrove roots that extend about three to four feet above the water level and water that ranges from three to eight feet deep, depending on the tide and location.

There were no roads that went all the way from the canal zone to Colombia. There were isolated bits of high ground and some game trails that the Indian tribes used for hunting, but even the Indians were not enthusiastic about going there.

It's steaming hot, and the air is so thick with water that it's hard to breathe. The jungle teems with mosquitoes, snakes, insects, jaguars, ocelots, and caiman. A caiman is like an alligator but stands up on all four legs and runs like a deer. You don't want to run into any of these critters, but if you go in there, you will.

On the Atlantic side, the Kuna Indians, known as the San Blas to the Spanish-speaking people in Panama, have their villages on small islands off the coast. It's a matriarchal society, and the women run the village. Their greatest punishment for criminal actions was banishment to the mainland. That tells you a lot about what they thought about the jungle.

We would reach the village by flying out by helicopter to a small patch of high ground on the coast where the Panamanian Coast Guard had their base and travel in a dugout canoe to the village. Beautiful blue water and quite a trip, but you didn't want to make the run at night.

Great white sharks abound, and Colombian drug runners make their runs at night in those waters. The sharks come there to breed in the brackish waters where fresh and saltwater mix.

Colombians come there because very few people live there. It's a good place to pass their cargo from Colombian boats to Panamanian fishing boats to smuggle into the country. Lots of hanky panky going on.

Our guys worked with the villagers and the Panamanian Coast Guard. There was not much trouble with Noriegistas there, but the Colombian drug runners were a threat that required some outside observation to assess the level of law enforcement involvement. That meant living with the Coast Guard and gaining the respect and confidence of the Indians as a second point of reference.

We never visited there without paying our respects to the chief at the island village. She appreciated the attention but didn't want gringos, or any other strangers in her village after dark.

On the Pacific Coast, we had the Darien Province safe house. If anything, this site was even more remote and rugged. We requested a helicopter for this one but were told to use the Panamanian commercial flight that ran down there once a week.

The new company commander wanted the ODA commanders to come to Amador to pick up their operations funds each month and brief him on what was happening in their sectors. It seemed like a good idea, and it would save me about a week each month of running across the country. I didn't mind, but it took me away from other things I needed to take care of on Amador, so I sent the message to the ODA commanders.

But the ODA commander in the Darien site begged us to reconsider. He was dead set against making that flight every month. He told us it had a fifty percent success rate. He had made the flight himself six times, and they had been forced to make emergency landings three times along the way.

At first I thought he was exaggerating, so I flew down there myself. They were due for a shipment of money, and I hadn't made that run, so this would be a good chance to assess the situation.

Since I was back in Amador I was back in uniform. The jungle sites could be in civilian clothes if they wanted to, but there were no reporters down there and our jungle fatigues made more sense for daily wear. I drew the funds, but this time I would be going in jungle fatigues and carrying my M16 rifle and pistol on my load-bearing equipment.

So there I was, sitting on a twin-engine Cessna at Tocumen Airport waiting to take off for the Darien. The engines were running and suddenly the stewardess opened the aircraft door and ran out.

That's strange.

Then, the co-pilot runs out the open door. The engines are still running, propellers turning.

That's very strange.

Then, the pilot runs off the aircraft carrying a fire extinguisher. The entire flight crew has now left the aircraft with the engines running.

That's too strange; it's time to see what's going on here.

I unbuckle my seatbelt and move toward the open door. Outside, I see the stewardess, pilot, and copilot throwing dirt on the portside engine, which is pouring oil and, by the way, is on fire.

And still running, prop turning.

The now empty fire extinguisher is lying on the grass nearby, and the flight crew is running back and forth to the edge of the flight strip to get loose dirt to throw onto the blazing engine.

Still running, prop turning.

Finally, the oil flow slows and then stops. The fire sputters out. The prop is still turning, and the crew looks at the pilot, who shouts, "Bueno!"

And they shoo me back inside, get on board, and we take off.

OK, maybe the ODA commander down there has a point. I spent most of the flight watching that left-hand engine.

The Darien is everything I expected. The main street of the town is eight-inch-thick mud. It's not only the main street; it's the only street. I chuckle as I watch a U.S. government official who flew down with us head up the street in her high heels to the town's one hotel clutching her high heels she had to dig out of the mud. She was in for quite a surprise when she got to the hotel. Let's just say I don't recommend it for your next vacation, and I strongly recommend you poke the mattress on the bed with a stick before you get close to it.

I get the town tour from the ODA, which takes about ten minutes. There are only twenty buildings. The jungle runs up to the edge of town. It is best not to be out after dark; the jaguars and other animals sometimes come in looking for food. The rats alone are the size of a cocker spaniel, and they would eat a cocker spaniel if they found one in that town.

I finish my work there and get ready to fly back. This will be even more of a white-knuckle flight than the one down. I watch the pilot and co-pilot talking to the owner of the dirt flight strip near his palm-thatched "control tower." The "tower" is about ten feet tall, made of logs with a palm thatch roof, and is nothing more than an elevated open-air platform with a ladder.

The pilot and the flight strip owner reach an agreement, and the flight strip owner tells two kids to run to the end of the runway and move a palm tree log lying across the runway. There are stakes in the ground every ten meters on both sides of the flight strip. The kids carefully place the log between stakes to block the runway about seventy-five meters away.

And calling this a runway is being generous. The local kids try to keep the dirt strip of hard-packed clay fairly flat and level by filling in the potholes after the daily rain. But there are marks on the palm trees lining the flight strip where planes have slid off it when it's wet.

Turns out the pilot has to rent the runway every time he takes off or lands. Landing is a set rate, but he has to pay by the meter for the runway used to take off. He has negotiated for seventy-five meters of runway for takeoff.

We board, and the pilot revs the engines to maximum. He and the flight crew cross themselves and pray out loud (and I mean really loud). Then, he pops the brakes, and we take off, barely clearing the palm log at the end of the runway.

OK, I get it. That op fund I just delivered will last a few months, and we can excuse them from the monthly meeting.

We go through a relatively normal couple of months with some tense moments, but the coup attempts appear to be dying down. But the bad guys are still out there. It's a waiting game. They know we will leave eventually, so they are laying low.

There are some attempts to identify the ring leaders, and we have arrested a few, but we have no definitive proof. Our mission is extended once, twice, and then three times. The Panamanian and U.S. governments are nervous about ending the mission, knowing what will happen when we go.

Finally, we received the order to pull everyone out of the site and back into Amador. But we didn't leave the country; we set up bunks in Noriega's old headquarters and waited. We won't have to wait very long.

The story of the team's recovery from Darien is worth telling here. I flew back down there again, this time in a Blackhawk helicopter, to recover the team and all their gear.

As we load the aircraft, I see the team medic talking earnestly to his team leader. His team leader keeps shaking his head no, but the medic persists.

Finally, the team leader shrugs and points to me. Ask him, if he says it's OK, I don't care. I know this gesture well.

The medic comes up and presents his case.

"Sir, you've got to let us take Kitty back with us."

"Kitty? This is about a pet cat? Are you kidding me?"

The team leader, laughing, tells me, "This isn't your normal house cat. Come on, I'll show you."

They take me back into the team house, the medic hands me a baby formula bottle, and then he hands me a Kitty. Kitty is a Margay.

A Margay is a wild cat similar to an Ocelot. It is slightly smaller but still a wild cat. At that time, it was an endangered species. No way, Jose. Put it back where you found it.

"We can't, sir. We found it on a patrol with the Panamanians out in the jungle. Hunters had killed its Mom, and they were going to kill it, too. We rescued it. It's just a baby; it won't survive without us. If we can return it to the Canal Zone, we will give it to the zoo at the Jungle Warfare Course. The NCOIC has already agreed to take it. He will meet us at the LZ."

I am sitting here feeding a wild animal with a baby bottle and watching a big, burly green beret medic with tears in his eyes begging for Kitty's life.

"OK, but it goes straight to the zoo when we get back, got it?"

"Roger that, sir, thanks a million!"

So we boarded the helicopter with Kitty for the flight back. But Kitty doesn't like loud helicopters. Kitty grows half-inch claws and teeth you wouldn't expect on a little kitten, and our SF medic is about to become his own patient.

Kitty gets placed into a nylon mesh laundry bag and set on the floor. But as the helicopter lifts off, Kitty likes this even less. Kitty tries to climb up the legs of the team members near him, bag and all.

Kitty rides back to Amador in the nylon bag tied to the barrel of an M16, held out the door and away from everyone's legs. Not a pleasant flight for anyone.

As we land, I hope to see the NCOIC from the JOTC course, but instead, I see our company commander, the SOUTHCOM commander, the mayor of Panama City, and a whole raft of reporters and photographers.

I am going to jail.

I can see the headlines now. "Green beret attempts to smuggle an endangered species and is captured red-handed."

It turns out this is the last team from the mission to be brought in, and it's being recorded for posterity. All the big wheels are here.

How do I get myself into these things?

The reporters start taking photos as we leave the bird with Kitty still in the laundry bag. I can see my boss's eyes widen, and I'm desperately trying to think of an excuse, but I'm coming up empty.

The mayor then comes forward and starts gushing, "A Margay! And a female, too! This is wonderful. We have a male in our zoo in the city, and now we can breed them here. You know they are almost extinct? This is a great day for Panama. Thank you, thank you!"

Whew.

My boss is grinning at me. I still have a lot of explaining to do, but at least it won't be through jail bars.

With the recovery of the Darien ODA, everyone was back at Amador. And things are heating up.

We began getting reports from Panamanians loyal to the new government that the Noriegistas were getting bold again. They are openly ignoring their leaders who are loyal to the government. Cuartels that have leaders who are not loyal deny any problems, but we notice they are in their offices when we call, even late at night. They all own large homes off the cuartel and being in the office late at night is not normal behavior. They are in there waiting for a call but not from us.

Things come to a head when Colonel Herrera, the head of the PNP, is arrested and placed in jail on the Amador peninsula. Amador was Noriega's personal headquarters, including his residence, offices, and a jail on the end of a peninsula for Noriega's "special" prisoners. There is even a house for Noriega's witch doctor on the same strip of land leading to the jail.

The jail was very secure, and we were on the other end of the peninsula on the mainland. It's about as secure a location as the Panamanian government can find. Herrera was brought back to Panama from exile as an anti-Noriega National Guard leader to take over the new PNP. But the series of explosions in David and other issues placed his loyalty in question.

This was strange because Herrera was considered one of the "good guys." Whatever the reason, he was held accountable for the unrest. After his firing, he was accused of being involved with the conspirators, which led to his arrest.

But on December 4th, 1990, a PNP helicopter flew over Amador and fired a burst from a machine gun at us on the compound. No one was hit, but the helicopter flew out to the prison, landed, and Herrera ran out from the high-security facility, got on board, and was gone.

Naturally, none of the PNP guards at the facility had any idea how he got out of his cell. But he was out; there was no doubt of that.

That night we started working the telephone to call all the cuartels and remind them who their friends were. Every senior officer is at his desk at ten o'clock at night, and they pick up on the first ring.

The police leaders out in the countryside mostly hedge their bets until they get a sense of who will win this one. Some are solid, and we know they are behind the government. Some claim they are, and we know they will roll over if it looks like a coup can succeed.

As the night goes on, we get reports of Herrera gathering forces. A cuartel near Colon calls in to let us know Herra and a group of men held them at gunpoint and took all the weapons from their armory.

Herrera moves and gathers troublemakers to him as he goes. We get more reports throughout the day and into the next night.

On the night of 5 December, we went down to the main police station to get an update on what was happening in the city. Shortly after we returned, we had a leadership meeting of all the team leaders when the radio, our main link to two of our SF guys working with the police in the PNP headquarters, lit up.

"Romeo six this Papa one niner. We have a problem."

"Roger that Papa one niner, this is Romeo six. What's your problem?"

"We have men with automatic weapons and body armor coming in through the doors and windows and taking over the building."

OK, that qualifies as a problem. The boss keeps talking to our two guys while I get the company up and ready for action. Looks like the coup is on, and there is no telling what will happen next, so it is best to ensure our site is secure and we are ready to fight.

We kit up and establish a perimeter around the building. We are on a base guarded by American MPs, with a fence and a gate, and our compound is inside another high-security fence, but there is no sense in taking chances. We get an ODA's worth of guys armed and outside to keep an eye on approaches.

We send the company sergeant major and a few guys to the PNP office on base to check on them and remind them whose side they are on. We leave a few guys with them just in case they forget.

We now have the base secured, so I checked back with the boss for instructions. Back at PNP Head Quarters, our two guys are on the top floor, and Herrera and a hundred or so guys with AK47s and other assorted weapons have taken over the ground floor.

The company commander is talking to our guys on the radio. Both of the SF NCOs are at the top of the stairs with pistols drawn, pointing their pistols at the hundred-plus Panamanians on the ground floor, who are pointing automatic weapons up at them. Herrera didn't want to kill Americans, but he needed control of the building. He was ordering them to put down their weapons.

But our guys stood firm.

The company commander wants to talk to Herrera and calls a phone number downstairs. He gets Herrera on the line, and they talk. Meanwhile, I get the company kitted up and ready to go.

Those are our guys. There are a hundred Panamanians in there, but there are seventy of us. If we go in, it will be bloody, but we are not leaving our guys to go it alone. The company commander explains this to Herrera as we kit up.

We are ready to roll, but the commander calls the SOUTHCOM commander to let him know we are ready, and SOUTHCOM shuts us down. No one wants a blood bath, so we need to try to talk this down. Be ready, but don't go. In the meantime, SOUTHCOM conventional forces are setting up a cordon around the building.

Things are getting more tense by the minute. The Panamanians tend to sit on the fence and can be pushed either way. If they take our guys on the top floor, they have American hostages, and this is going to change the dynamics of this situation. In a way, they already have them hostage since they can't come out, and if we go in the shooting will start and they will probably be the first ones killed. I need something to tilt the balance our way. And I've got an idea.

I called Howard Air Force Base nearby. "Guys, I need a C130 in the air and doing counterclockwise race tracks low over the PNP Headquarters."

"We don't have any gunships; they are all in the desert." Operation Desert Shield was up and running, and all assets were there.

"It doesn't matter; they won't be shooting at anyone. I just need the guys inside to hear that bird circling overhead. It's dark, and they can't see what kind of bird it is."

"OK, got it, we will scramble a crew right now. Give us half an hour, and we will be on station."

We keep talking to Herrera, and SOUTHCOM also talks to him. Finally, Herrera knows he has a stalemate and he needs to up the ante. He calls our company commander and tells him, "I don't want to hurt your men, but they need to lay down their guns now and surrender, or they will be harmed." Herrera is gambling on getting them alive; this will make a good bargaining chip for him.

I got a call from Howard AFB, and the bird is now on station. I advised the company commander and suggested his conversation with Herrera. The boss grins and picks up the phone.

"Colonel Herrera, stick your head out the window and tell me what you hear."

"What?"

"Just listen closely and tell me what you hear above you."

There is a moment of silence, and then Herrera returns to the line.

"If you have that damn AC130 fire on us, you will kill your men. You need to think about that." Herrera is a little rattled now. The AC130 was a key weapon during the invasion and is greatly feared. It decimated PDF forces in more than one location, including a police building right in the city not far from where we are.

"Yes, my Colonel, you are correct. If we fire on the building, it will kill my men on the top floor. And all of you on the bottom floor. As long as I can talk to my men on the top floor and they are unharmed, I won't destroy the whole damn building. I suggest YOU think about THAT."

We pass the deal on to our guys standing at the top of that staircase. That gives them a chuckle.

Checkmate. Whatcha gonna do now, Ranger?

Herrera ponders this situation, but he is now at a dead end. He needs to change the equation again. The Panamanians inside use vehicles to ram the fence around the compound and break out on a side not expected, trying to dash to the Presidential Palace. They won't make it past the cordon. One is killed; the rest captured. Herrera escapes but is captured shortly afterward.

And that was it. The hard cases had thrown the dice and lost. The coup was dead, and there would be no more attempts to bring back the power of the PDF. The plan to draw them out had worked, and Panama now had a chance to build a new future.

Our work was done, and we would soon re-deploy back to the U.S.

Chapter Twelve

USASFC

When we got back from Panama, I was reassigned to the new U.S. Army Special Forces Command (USASFC) Training and Evaluation Branch. This new headquarters was searching for a mission, and the process was painful for all concerned. The guys on the ODAs called it "Useless Fuck." It's not exactly where I wanted to be.

The official mission was to train, equip, and manage the groups. It was not a combatant command. But it had a general officer in charge, and general officers tended to want to command. As a result, there were growing pains.

The branch I was assigned to was responsible for exactly what the name said, Training Evaluation. Their mission was to test ODAs before they deployed on tasks the command thought were important. In theory, no ODA could deploy until and unless they were "certified" as trained by our headquarters.

It sounds good on paper, but the devil is in the details. SF was a busy place in those days. Teams were constantly deployed to Colombia, Panama, and now the Middle East. We had entered a new era, and SF was in demand worldwide.

This meant back-to-back deployments. And on a deployment, you often didn't have time to train on anything not directly related to the mission. Many missions had the team split up into small groups. Rifle and pistol marksmanship, in some cases, was restricted to shooting back at the guys who were shooting at you and keeping your weapon concealed the rest of the time.

However, certification was based on classic SF missions and tasks, mainly around easily tested tasks. Rifle and pistol marksmanship was one, and twenty-mile road marches were another.

As a result, teams that came back from deployment had to certify before they could redeploy. But if you hadn't been doing long road marches, your feet were not hardened up and you suffered.

The original certification test was five twenty-mile road marches, one each day for five days straight, with rifle and pistol qualification at the end of the day and other tests. It was a serious gut check you needed to train to accomplish without trashing your feet.

We were also shorthanded, so much so that Congress passed a law stating that we could not be deployed more than 180 days straight before we came home so the families would get a break. But the missions kept coming.

To meet the demand, the groups started bringing teams back from deployment only to get a new set of orders and go right back out. You would be home for a week or two, then redeploy. I got that cycle with the Colombia and Panama missions.

But now, for the guys coming home to get a break, certification hit right in the middle. For one of their two weeks at home, the guys would have to go out and go through this meat grinder just to prove they were qualified to go back and do what they had been doing to begin with. And you didn't get to go home. You road marched between points and camped on the ranges for a week.

As a result, a well-intentioned law to give families time together simply became a dreaded, major hassle. Needless to say, it wasn't popular among the teams. Also, of all the assignments I could have gone to, this is the last one I wanted. As my team said to me when I was leaving, "Well, Cap, at least now the enemy will have a face."

The only saving grace was that the pain of certification was beginning to subside by the time I got there. Common sense was beginning to kick in, or more likely, the damage certification was causing to the troop's morale, as well as the families, was interfering with the group's ability to get teams back out the door to meet mission requirements. Certification was settling into a more reasonable routine.

I was assigned to set up a certification program for National Guard individual replacements being sent to Desert Shield. Desert Storm had not started yet, but it was coming.

Instead of the usual ball-buster routines, I worked hard to incorporate actual training into the program. I got the guys designated to go out as individual replacements for three weeks, so we had more time than normal.

I set up the physical events at the end of the three weeks and scheduled daily physical training and road marches that would build up to a single twenty-miler at the end. We were now down to one twenty-mile road march in six hours or less. That is still a ball buster, but it's doable as long as you train. In between, we went to the range several times for qualification and short-range rapid engagement drills.

Instead of just rifle and pistol qualification, I set up training and qualification on every weapons system they might encounter during deployment. Medium and heavy machine guns, NATO and Warsaw Pact weapons, anti-tank weapons, sniper rifles, and even Stinger anti-aircraft weapons.

Some of the weapons were brand new and just being fielded. I found out later that as Desert Storm started, at least one ODA was issued a sniper weapon system they had never seen before. As luck would have it, one of my guys came in on the truck with it. He could put it into action right off the truck, and it was needed. I felt like I had accomplished something.

I also was tasked to go out and walk with ODAs when they were training, something I enjoyed for the most part.

One assignment I drew in the dead of winter was walking with the 10th Special Forces Group at Fort Devons, Massachusetts, during their annual winter exercise. It sounded fun, so I asked permission to draw extreme cold weather gear for the trip.

The problem was that we didn't have that gear at Fort Bragg. I was told the 10th SFG would issue me a set when I arrived, but as always, I took what I had, not assuming anything.

When I got to Fort Devons, I went down to the Group CIF for a cold weather equipment issue and was laughed out of the building. "It's all issued out, sir; see for yourself." Sure enough, the cupboard was bare. And the guys there were none too worried about a hated USASFC command evaluator who didn't have the proper kit. I got a set of over-whites for camouflage, but that was it.

All I had was a good pair of leather boots with a little insulation, a set of Gore-Tex, some lightweight long underwear, a wool sweater, and a field jacket liner. Plus some gloves and a watch cap. Plenty for Fort Bragg, not so good for sub-zero weather in New England.

But I had served with less in Bad Tolz and figured I could handle it. After all, I was an evaluator and wouldn't have the same load as the ODA's. If it got too bad, I could ask to be pulled out. No way was I going to tell these guys I couldn't go out with them because I might get cold.

So off we went. As always seemed to be my fate, as we were gearing up to go a major North Easter came blowing in, with snow, strong winds, and sub-zero weather. The jump-in was not possible; we would have to truck in.

Everyone was grinning at the USASFC staff guy as we got on the truck, but still, I wasn't too worried. I had been in this kind of fix before; if I was careful and followed the cold weather rules of survival, I should be fine.

We also had an evaluator from their Group headquarters with us. This poor ODA was overrun with staff officers who were there to see if they knew what they were doing. I figured a major part of this poor captain's job was keeping us staff pukes from getting hurt, so I was determined not to add to his burden any more than necessary.

The storm was in full blow when we got to the drop-off point. I've been in some pretty bad winter storms, from Bad Tolz to Ft Bragg to Alaska, but this one came close to taking first place. The wind was howling, and the scene looked like something out of the Doctor Zhivago movie.

We started moving cross country, and my fellow evaluator moved ahead on skis. He had the better equipment for this section, which was open fields. We were on snowshoes and he could move much faster. But we soon got to the wood line, and things changed quickly.

The skis were not maneuverable in the dense thickets, and he kept getting tangled up. Finally, as we moved near a small stream that was frozen over he lost his balance and went into the stream bed. And right through the ice.

As we pulled him out, the water was freezing on his Gore-Tex as soon as it hit the air. It was that cold. He was soaked and wouldn't last long in this storm.

There was no choice but to set up camp, get him out of those wet clothes and warm him up, which suited me just fine. I had been hanging in there, but my clothes could not handle the cold unless we kept moving.

The ODA was doing the right thing, making security halts every few minutes to look, listen, avoid detection and ensure the new snow and wind covered their path. I couldn't fault them for their tactics, but I wasn't equipped to crouch in the snow for long periods. I was headed for hypothermia and frostbite if we continued at that pace. I had better cold weather equipment than in Alaska, but I had no 300-pound Akio to pull to keep me warm. My fellow evaluator hit the water just in time.

The ODA did fine; the storm passed, and I got through my field trip without issues. There were other trips to the field like this one, but overall, I was doing staff work, which was not greatly appreciated.

At least by the ODAs I wanted to work with. For most SF captains at that time, once you left an ODA, that was it. You would never serve on an ODA again. Less than five percent of SF officers got a second ODA command. But I was determined to be one of that five percent.

The good thing about staff work was that I had time with my son. The divorce and my being gone on missions hadn't been easy on him. Being home was important, and I tried to make the best use of the time. We joined the Cub Scouts, him as a Cub and me as an assistant Den leader, made some trips back south to Alabama to see my mother and his grandmother, and had some good times.

But after two years on staff, it was time for me to be re-assigned, and I was due for an overseas tour as an officer. I faced another difficult choice: Going on an overseas assignment as a single parent would be tough for me and my son. The forward-deployed battalions were constantly deployed on missions you simply do not talk about. Who was going to look after my son when I was out?

An overseas staff position seemed to be my only option. I wasn't looking forward to that, but what choice did I have? Then two things happened. First, some close friends who had looked after my son when I had to go out on missions from Ft Bragg told me they had orders for Kadena Air Force Base in Okinawa, Japan. As luck would have it, a forward-deployed Special Forces battalion was on Okinawa.

But my USASFC boss told me I had been selected to be the Head Quarters Company commander. That job was probably one of the worst you can get as a captain. You are the commander of a company where nearly everyone there outranks you. All the senior officers in the command are assigned to that company: Generals, Colonels, Majors, etc. You have to ensure they take their physical fitness test on time, get qualified with their weapons, and do all the regular housekeeping duties of running a company.

You also have to set up all the social functions and parades, you name it. In short, it sucks. Nothing remotely sexy about it. But you work for the General and are around all the senior officers in SF. They get to know you, and it's a pretty good way to ensure your next couple of promotions.

This also pre-empted an overseas assignment if I chose to take the job. I would have to go overseas after my tour as a company commander in USASFC, and it would also set me up for a second company command overseas as a Major. But I would never serve on an ODA again.

The command felt they were doing me a huge favor, so they were more than a little shocked when I turned it down. Yes, I know this will be a huge boost to my career. And no, I don't want it or the promotions that may follow.

I wanted another ODA.

In Okinawa.

And so I got one.

Chapter Thirteen

Behind the Curtain

My next ODA was going to be an interesting assignment. We were to go places no one else had gone and assess the situation for others who may follow.

This meant keeping a low profile for much of our work. It's not a James Bond thing; no cover stories or spy tradecraft were employed. It simply meant wearing civilian clothes and keeping your mouth shut about who you were and what you were doing. Much of what we needed to do was in very bad places where very bad people were doing very bad things.

For us and anyone else who might follow us later, a big part of the job was researching how to get in and back out again alive. This was mostly done with the host nation's knowledge, if not their support. We would work in small groups, sometimes alone.

As I mentioned, we didn't use cover stories. If asked, we would tell the truth, to an extent. We were Department of Defense employees who were there to work on the embassy's Non-Combatant Evacuation (NEO) plan. We wore suits and let our hair grow a little longer. Most who saw us thought we were Embassy employees; we were considered tourists on the streets.

You would be amazed at how many tourists there are, even in the most dangerous places. It's not unusual for some of them to wind up missing or dead, but there are still more who pop up despite the danger. Some businessmen are there despite the risk in pursuit of profit. Others were "adventure tourists" naive enough to think that they would be safe since they were nice people. We simply blended in with whomever we were among and let people believe what they wanted to believe.

If questioned by officials of the nation we were in, we would identify ourselves as being with the U.S. embassy, which was true. Sometimes, they even assisted us with what we needed for the NEO planning. Sometimes not.

Most of what we needed was accurate information on routes to and from the airport and seaports. The capital cities of many nations, especially in that part of the world, are densely packed with unmarked streets that do not follow any logical pattern. It's easy to get lost, and there are plenty of places you don't want to be lost in.

We would learn how the local people felt about Americans along the way. It wasn't always welcoming. Understanding how to move through slums and dangerous places, where you stand out and lack armed security, is key to success and staying alive. Identifying those dangerous areas and the best path through them is part of the job.

We were on one job in one of those dangerous areas after the U.S. had bombed Libya, and the two nations were not on friendly terms. We were out and about mapping out a route using a military GPS.

I didn't like using the GPS. It was big, about the size of a cigar box, OD green, and military. It was one of those "signatures" you don't want to wave around and draw attention to. Usually, we would keep it inside the car and out of sight. At the time, it was state-of-the-art and all we had. Commercial GPS was not yet a reality, at least at the level of detail we needed.

One of its many drawbacks was that it didn't work very well when you were moving. It was an early model, and back then, you needed to be stationary for a few minutes to let it find the satellites and get a good reading. We didn't have that many satellites in those days, and the systems that let you navigate on the move hadn't been fully developed. But accurate GPS readings were important so we usually stopped near key intersections to get a good reading.

We were at one such intersection, next to a compound with a high whitewashed wall. It was the only place to stop that wasn't in the middle of crowds of people and goats. It was near a large tree, giving us some cover from casual observation. It looked fairly inconspicuous, and we only planned to be there momentarily.

My guy with the GPS had a reputation for being a little anal about his job. He would go for the 100 percent solution when the 90 percent would do just fine. I have nothing against the pursuit of perfection, but sometimes you have to weigh risk vs accuracy and decide in favor of staying alive.

My guy hadn't brought the car top antenna and struggled to get a good reading inside the car. I allowed him to get out of the car and lay the GPS on the roof, but inside his backpack, while he got a better reading. But only for a moment.

One minute turned into two, then three, then five. I told my guy to get back in the car; this was taking too long. He resisted, giving me the "just a minute, it's almost there." Finally, I had to get out of the car myself to quietly tell him to knock it off, do what I told him, and get back in the car. It was not a place to be shouting orders and drawing attention, but we needed to move.

As I got out, I immediately noticed two things. First, a CCTV camera I couldn't see from inside the car near the tree that had been covered by a branch. It was aimed right at where the car was parked next to the wall. They had a clear shot of my guy getting a reading on this location with a military GPS. Not good.

Worse, beyond the large tree, I could see a very large man with a large lump under his shirt on his left side. And his right hand inside his shirt near the large lump. He was leaning out from a door in the wall, looking towards us nervously but not approaching.

I quietly closed my guy's book bag and told him to get his ass back in the car now and not argue with me. SF guys are notoriously independent and prone to argue about almost anything, but they know when an order is an order and the line has been crossed. We both got into the car and pulled off.

As we did, we could see the large brass plaque on the wall hidden by the tree.

"Embassy of the People's Republic of Libya."

OK, I guess I can see why they would be a little nervous about four Americans with a military GPS taking a reading right outside their embassy.

Nothing personal, guys; you just happened to be on the route. But we will ensure you get mentioned.

We encountered many situations like that one. They were usually easily avoided, although things would get a little too hot at times. Our mission was to get in, get what was needed, and get back out without incident. We were usually followed by the host nation's secret police, which we didn't mind. They were briefed on what we were doing, and it was no secret to them. Their presence helped ensure we weren't picked up by someone less friendly.

But occasionally, we picked up more than one tail. Spotting a tail is an art form that is not easily learned but is essential to survival.

Most of us had previously spent time in bad places and knew the signs. It was one of the prerequisites to getting on this team. You usually don't spot a tail right away. When you spot them, it's only because you have seen them before, usually at least three times. There are several ways to spot a tail, but it all boils down to repetition and something that doesn't fit.

Seeing the same guy or car three times, especially if you are taking a random route to see if someone is following you, is one technique. Most good intelligence services don't make that mistake. They use teams of people and rotate them so you never see the same person or car two times in a row.

But everyone has limited budgets, and sooner or later, they will rotate the same car or person back onto your trail. You must constantly scan the crowd and remember details to spot the repetition. Sometimes, it's days before you realize you have seen that guy across the street one too many times.

The second giveaway is the use of tradecraft. This involves the guy or girl following you and making a mistake you spot. Tradecraft takes many forms. Sometimes, the people following you don't have enough people to rotate on your tail. So they carry different color jackets, hats, wigs, anything to change their appearance and look like someone other than who they are.

If you are alert and watch the details, you can sometimes spot the guy with the same height, weight, skin color, and build you saw a moment ago, who suddenly has a new jacket, hat, and sunglasses. Dead giveaway.

You can quickly change hats and jackets from a large purse or shopping bag or buy one while passing through a store, but changing shoes and trousers is harder. Those telltales let you know this is someone you need to pay attention to, especially if the shoes are of a type no one else who lives there normally wears.

Terrorist groups normally don't have the resources of national intelligence services, and they are the most likely to use the same people to follow you around. Usually, they don't follow; they simply stake out where you live, your hotel, or where you work. The flower vendor is across the street from the hotel, for instance. Especially if there is a florist inside the hotel, and that guy across the street hasn't sold a flower in days but still comes there daily.

Beggars on the same street corner every day and people loitering where they have no reason to, especially ones with cameras who take photos or videos when you pass by, are all red flags. And when you see the same guy or a couple of guys everywhere you go who are not professional intelligence service but are keeping tabs on you, it's time to get worried. Especially if they quickly break off and disappear when the real intelligence service tail shows up.

We would get that every once in a while. Several ways to handle that. One is to notify the Regional Security Officer at the embassy and let him notify the host nation's intelligence services. If it's their guys you spotted, no harm done. Sort of a "tag, you're it" notice. It's slightly embarrassing, but they won't admit it's them; they will just adjust their tactics or personnel.

If it's not them, they are the ones who need to know. When they start getting worried, it's time for you to move. New hotel, different town, or different country. The trick is to stay one step ahead.

Terrorist groups often use amateurs for their surveillance, people who receive pay for information but are not part of their organization. Surveillance is a high-risk job; if the local is spotted and captured, they cannot tell the police much about their bosses. It helps us since they are not trained and are easier to spot, and the terrorists can always get more to work for them.

But when they start popping up, you are inside the terrorist operational planning cycle. It starts with surveillance, and it ends with you dead or sitting in orange pajamas in front of a video camera with a guy holding a big knife standing behind you. It's a cycle you want to disrupt.

We would encounter that several times, usually ending with us simply leaving the country. The easiest way to break the cycle without incident. We could always come back.

Our biggest mission turned out not to be our mission at all. During President Clinton's administration, the "truce" in Korea was getting very strained. At one time, over twenty intelligence indicators were used to determine if the North Koreans would invade. They hit all of them and took other actions that no one had considered.

At that time, North Korea had not tested a nuclear device, but we knew they were working on one, and they might have completed it. That could be part of their calculus for raising the threat of invasion.

The capital in Seoul is only an hour's drive from the border, and we knew we couldn't stop them from getting into Seoul if they came. Getting them out again would be tough, especially if they brought a nuke.

But they didn't quite cross the line to all-out invasion. There was a North Korean submarine that was infiltrating Special Forces onto the beach in South Korea (big indicator) that got itself beached and caught. The North Korean Special Forces team led everyone on a merry chase for weeks while the South Koreans and U.S. Special Operations tried to catch them. That one resulted in several hot engagements in South Korea and casualties on both sides. The North Korean commander was the only one who survived, and the last we saw of him was boot prints going into the minefield on the DMZ and heading north.

There were other incidents as well, and the prevailing thought was that war with North Korea was imminent. President Clinton was concerned enough to warn the North Koreans that if they had used a nuclear device on the south, "North Korea would cease to exist as a nation." But as the Commander in Chief of U.S. Forces in Korea said at the time, "If anyone tells you with any degree of assurance what the North Koreans will or will not do, never listen to him again. He knows nothing about these people."

We specialized in helping people escape bad situations, so we were assigned to South Korea to examine their NEO plan and offer advice.

Now, this was a very bad idea for several reasons. First, we were a sensitive, classified unit operating under the radar. It's not undercover; it's just a low profile. SF no longer performs that mission, but it was a closely guarded secret at the time. Our success depended on not attracting attention and not being openly identified as military, let alone Special Forces. But in Korea, we would be immersed in U.S. Army personnel, many of whom knew us personally. It would be very difficult to hide.

Second, our specialty was finding a route through bad places so a small group of highly trained operators could get a small group of U.S. personnel out if needed. In Korea, we were talking about thousands spread across the country. This was a logistical nightmare, and officers were better trained and more experienced than we had studied for a long time. It was very unlikely that we would have anything useful to offer to this plan, and worse, we would be "grading the paper" of general officers above us. It was not likely to win us or Special Forces' many friends.

I spoke out against this one and tried to decline the mission. It was the first and only time I ever tried to do that, but it went nowhere. We had handlers at the Commander in Chief Pacific

Command (CINCPAC) we worked for when we were out on these missions who had already volunteered their assistance, and we were their only resource. It was a bad idea, but we were going anyway.

I found myself as an Army captain with a non-regulation haircut and a suit sitting in the US embassy in Seoul, getting their briefing on the NEO plan for the Republic of Korea.

Two significant facts jumped out immediately. First, the U.S. Army assembled and evacuated the personnel covered under this plan. No surprise there, but we would have to meet with the military officers responsible for this plan to get into any details. That was exactly what I didn't want and precisely what the embassy had already arranged.

We would be briefed the next day by what the U.S. Forces Korea command considered to be "an officer of appropriate rank" to brief the "diplomats" from the U.S. Embassy on the NEO plan. A two-star General. The "diplomats" would be me and a Sergeant First Class from my team in civilian suits, but I sincerely hoped that didn't come up in the conversation.

The second significant fact that came out in the embassy briefing was that the U.S. Embassy had committed to including all the staff of allied embassies in the U.S. NEO plan and their extended families. Their rough estimate was that approximately fifteen thousand people would require evacuation during a war.

The U.S. Embassy had not informed the military of this tidbit of information, and they also didn't want us to tell them. They figured the entire U.S. Army wouldn't have any trouble with this since they had plenty of resources, and "they would do what the ambassador told them to do." Typical Department of State planning.

Oddly enough, I started breathing a little easier right there. This was so blatantly stupid and poorly thought through that right there I could see where we could help out. All we had to do was convince the U.S. Embassy that their plan wouldn't work without telling the Generals that this was what they were up to.

I know—it sounds a lot tougher than it was—but the one thing that came ringing through was that the embassy hadn't discussed this with the military, and they had no idea what would happen in an invasion. We needed to get that brief for them and come back armed with facts that would show them what they had gotten themselves into with their smug "the Army will do what they are told to do by the ambassador" plan.

Piece of cake.

Of course, we still had to devise an alternate plan to get the embassy's fat out of the fire without losing too much face. That's a little harder, but hey, we are SF. We do the impossible for PT every morning.

So, there we were, an SF captain and SFC in our best suits and long hair, getting a briefing from a two-star General on the U.S. NEO plan for the Republic of Korea. Most of the briefing I spent looking around, praying no one was there I knew or would ever meet again.

My good sergeant, of course, was in his element, telling the full Colonel tasked to keep our teacups full to "bring some more of those little cookies" and calling out to the two-star "General, go over those figures for totals for evacuation your plan is based on one more time, and where did you get those numbers?"

The good news was we were left with a briefcase full of hard facts on what was available to support a NEO operation if an invasion started and how the battle would quickly evolve.

Long story short, the Generals based their plans on the assumption that if the entire North Korean army came over the border, they would be allowed to use the U.S. Army to try and stop them instead of obediently standing by for the U.S. Ambassador's instructions. And they didn't have the resources to conduct a concurrent NEO evacuation. It was going to be chaos, and the families and other non-combatants would be caught right in the middle. Early return of families before hostilities began and stopping new arrivals of non-combatants were the only viable options that might help.

The problem was clear. The requirement to fight a screaming horde of invading North Koreans was the military's problem, and the embassy has already made it clear they felt they and NEO came first. However, time or resources would not be available, regardless of what the embassy wanted.

You would think the problem would be obvious, but the light bulb hadn't gone on in the embassy yet, and DOS never likes to admit it was oblivious to the most obvious facts of the situation.

To get the point across to the embassy, we would need something fact-based that no one could argue with.

Lucky for us, the military brief gave me what I needed.

The assembly area for the personnel being evacuated from the area around the capital was on the Yongsan Army base. It was logical, it was close to the embassy, it had helicopters and Military Police, and it was the best way to get everyone together to get them out quickly.

This also recognized that if an invasion came, the entire city would be chaotic, and the roads leading out of the city would be worse. Seoul is only an hour's drive from the DMZ. Even with all the defenses, the North Koreans would enter the city within a day or two, and the entire city was well within range of the heavy artillery and rocket launchers of the North Korean Army where they sat inside North Korea.

In short, the only hope anyone had for a NEO was to get to Yongsan as fast as possible and get air-evaced out—within hours, certainly within the first day. After that, everyone would be fighting in the streets. No one wanted to lose Seoul, but all the actual warriors knew it would be a hard fight inside the city, and we would probably be pushed out.

Armed with this information, I had to devise a way to convince the ambassador to change his plan. This would be a huge change, not something they could just erase from their existing plan and keep quiet about. The embassy had painted itself into a very dangerous corner.

Two days later, I was sitting in the embassy briefing room with the U.S. Ambassador sitting across from me, waiting for "our findings."

In a situation like this one, where you know everyone is there prepared to defend their position to death, you need to open up with something no one can argue about. Cold, hard facts that cannot be ignored or leave an opening for opponents to defend their cherished position.

I opened with their plan. "Sir, the most significant issue was the sheer volume of personnel you hope to move under this concept. The U.S. military simply will not have the resources to accommodate them."

"Nonsense. With all those troops and aircraft, you cannot tell me they cannot evacuate even this number of personnel. It will take some time, but this is a priority, and they will simply have to deal with it."

"Yes, sir, but unfortunately, time is the only resource you do not have. Allow me to review a few critical facts. First, your agreements involve evacuating approximately fifteen thousand personnel, correct?"

"Correct."

"The assembly area for these personnel is on Yongsan, here in the capital. There are three gates leading into Yongsan. If you will humor me, I would like to run a simple experiment to demonstrate how time is a critical factor."

I pulled my ID card out of my wallet and held it up.

"Now, sir, I am going to hand you my ID, and I would like you to look at the name and Social Security Number and confirm that it matches what's on this piece of paper."

I slid a sheet of paper across the table to him, with about twenty names and SSNs listed.

He frowned but took my ID, glanced down the list, and handed it back to me. "It appears that you are on the list, and the number is correct. What's your point?"

"Sir, it took you about fifteen seconds to accomplish that task. Not much time, but we must remember that we are talking about fifteen thousand names on a list. Granted, we could use a computer database to check, but I think we can agree that it would take a few seconds more to enter the data. We could issue everyone a swipe card, but we would still need to verify that the person with the card matches the photograph. I think we can all agree that fifteen seconds is the minimum time we will need to verify the identity of the person trying to enter the gate, regardless of our technique. Do you agree?"

"Yes, yes, I agree, but what's your point?"

"Well, sir, some simple math is in order now." I pushed a pocket calculator across the table to him.

"If you would, sir, enter these figures if you agree they are valid. First, fifteen thousand personnel are to be evacuated, correct?"

"Correct." He entered that number.

"OK, multiply that by fifteen seconds."

"Done."

"Divide by three gates available for entry."

"Done."

"And divide by sixty to obtain minutes required to get everyone past this first very basic task needed to ID everyone to be evacuated, and again by sixty to obtain the number of hours."

He sat there for a few moments, looking at the final figure. "About twenty hours." He was frowning now, clearly unhappy with what was becoming more obvious.

"Now, consider this. Three gates won't be open. Only one will be available to use for evacuees, and the others will be closed and locked. The Military police will no longer be in Yongsan. They will only have one squad of eleven men. Our actual processing time is more than sixty hours."

"How is that possible?"

"Their wartime mission. When the invasion begins, the North Koreans will be only one hour drive from the city, unopposed. Of course, all available forces will oppose them to prevent that, but that requires our forces to deploy out of Yongsan and head toward the DMZ to prepared positions. Any diversion of forces from that wartime mission simply ensures the North Koreans will get here faster. The war plan depends on those MPs available for traffic control to help move forces north."

The ambassador's frown deepened.

"In addition, sir, our experiment here was done in very comfortable settings. Consider fifteen thousand extended families lined up outside that one gate, frantic to get inside, with only eleven MPs to get them processed and settled into the assembly area, which is far too small for that many personnel. And factor in the fact that we are in artillery range of North Korea right now, shells will be falling on the city, and most of the people in the city will be outside those gates as well trying to get inside."

I had him. The whole room was so quiet you could hear a pin drop. Well over twenty senior state department officials were in that room, and this had never occurred to any of them.

And then the question. "What do you suggest?"

This was the hard part. It would get resistance, but there wasn't any other way.

"Sir, I recommend you immediately implement a three-phase plan to drop the number of personnel straining the system."

"I cannot renege on our promise to our allies. They remain in the plan."

I had expected that. "Nor would I suggest that. But clearly, we cannot move them all at once. We need to develop a plan that moves them in stages."

A senior staffer beside the ambassador piped up, thinking he had me. "But we know the invasion forces will likely reach the city within forty-eight hours. We need our personnel moved before then."

"That's the first point of my three-phase plan. Stop moving personnel into Korea right now. We expect an invasion, yet you are still doing business as usual, rotating families and staff and bringing in new personnel weekly, correct?"

"More or less, but we need to sustain operations." The staffer replied.

"Conflict occurs in phases, just as your plan to address that conflict must have phases. I assume you would have greatly reduced staff during the war, correct?"

"Yes, but that's going to be very inconvenient."

The ambassador cut him off at this point. He had not been getting very good advice, and he realized it. "War is inconvenient, and we will have to make adjustments. He is right; we need to start now, not later. What else, Mr. Brewer?"

"Second, you need to seriously consider authorizing an early return to the U.S. or the other home nations of families here now. I also suggest you consider trimming your staff to wartime levels right now."

The same staffer jumped back in at this. "Do you realize the diplomatic statement made when you reduce embassy functions, let alone authorizing the early return of families?!? The press will go crazy!"

"Yes, I do, and I do not presume to tell you when you must do this. Certainly, it will send a message to the North Koreans and the world. But if the message you want to send is that the US is serious and intends to resist an invasion, I would suggest that this provides you a powerful diplomatic tool you can use."

Everyone got quiet again on that one.

"Last, I suggest you use your warden system to notify personnel who will be evacuated and work with the military to identify alternate assembly areas. Rather than have everyone move to the Yongsan assembly area at the first sign of trouble, tell them to remain in their homes until called forward. By doing this, you can control the flow of personnel and use their homes, food, and water for support until the pipeline can accept them. Otherwise, you will have chaos in the assembly area, insufficient shelter, food, or water, and thousands desperate to board a limited number of aircraft.

New staffer now, "But what about the North Koreans? What if they get to our houses before we are called forward? Who is going to protect us?"

"Good point, but those closest to Yongsan will get called in first. Correct me if I am wrong, but many of you live on the south side of the river, correct?"

Lots of nodding heads.

"Well, that river is a major line of defense. If I were you, I would not want to go north of that river, toward the North Koreans, if I could stay in my apartment with that river and a significant number of defenders between me and them. Plus, do you encounter traffic when you come to work in the morning, trying to get across those bridges?"

More nodding heads and a few eyes rolling. Traffic is notorious in Seoul, especially when crossing the river.

"Think about what it will be like with artillery falling on the city. Also, consider that the entire city will run south across that bridge when you try to drive north."

Dead quiet again. Another little detail no one thought about.

"In conclusion, sir, there are three phases and a recommendation. First, stop or delay new arrivals. Yes, that sends a message but use that to your advantage in negotiations. Second, early return. A stronger message and another diplomatic tool at your disposal to provide a graduated response to North Korea's aggressive behavior. Third, put your warden system to work to establish new assembly points and a call-forward plan. I recommend you fully brief the military on your agreement and work with them to establish your new assembly points and call forward plan. It won't work unless you involve them in the planning from the outset."

And we were done.

A flurry of staff actions followed to absolve those responsible for this debacle from responsibility and claim credit for the brilliant new plan developed by Department of State professionals, ably assisted by CINPAC senior staff officers.

A month or two later, the "three-phase plan developed by the U.S. embassy in the Republic of Korea in concert with officers from CINCPAC staff" was announced, but we remained behind the curtain, no one the wiser.

Chapter Fourteen

Klingons

I spent four years in Okinawa, mostly in the same company. I was the battalion assistant S3 for a while, a company executive officer in B company for a short time, and commander of the "long hair" team. However, two years into my three-year tour, I was selected for promotion to major.

This can present a challenge for a small unit like this stationed overseas. There are only five Major slots in a Special Forces battalion, so there are limited options on what to do with a new Major.

First, I could stay where I was working as the assistant S3 at that time. I wouldn't pin on Major for another six months when my sequence number came up, and by then, I would be "getting short" and preparing to rotate back to the U.S. Two majors in the S3 shop would not be a big deal.

Or I could rotate into an authorized Major slot. As luck would have it, there were two companies where the current commander was due to rotate, one of them the company I had been assigned to before as a detachment commander. However, that company's primary mission involved much more than the long hair mission.

C Company was the CINC's "Immediate Reaction Force". (For those who served there, this is what DOD officials directed that the unit be called in this story via the Pre-Publication Review.) This means that if the CINC has an emergency, the IRF goes in.

Special Forces live and work in the theater, know many of the local military and law enforcement leaders, and speak the local languages.

Special Forces trains Close Quarter Battle, but finding places to train on Okinawa is challenging. And the "Emergency" mission is not your only mission.

There are other units that train for this mission full-time, but they are based in the U.S. and it's a long flight to get to the nations in PACOM. The bottom line here is that for "emergencies," you need someone readily available.

Hence the "IRF." However, the Klingons in black body armor would likely never be used, although we had to train constantly for that mission. Precision marksmanship and shooting while moving are perishable skills. If you are "off the gun" for over two weeks, the instinctive "muscle memory" begins slipping away, and mistakes happen. There is no room for mistakes inside a dark room with people shooting in close proximity to one another.

This is a challenge for anyone in charge of an "IRF" type unit. At that time, most of what SF did was maintain relations with foreign militaries and their governments by training foreign armies. It helps build positive relations and lets our guys have a close-up, first-hand look at what's happening in those countries.

For nations that are our allies, it helps make their military stronger. For those who are not that friendly, it gives us a chance to work with people who usually move up in their military to higher ranks. When you train together, they get to know you, and you get to know them. It's a good first step to building respect and understanding that can pay off in the long run. It also helps us to know who their future leaders are and how they think.

We combined these training and mission requirements using C Company to train foreign military personnel in Close Quarters Battle. This is a term used for combat in cities or buildings up close and personal, where it's important to get the bad guys but not hurt civilians or friendly personnel in close proximity to the bad guys. By proximity, I mean inches, not yards.

We selected countries requesting training assistance based on locations where we might have to go to execute an actual hostage rescue or Close Quarters Battle mission. When I was commander, we had one such mission in the Philippines at a time when U.S. forces were not allowed in the Philippines by their government. The U.S. had been asked to leave, closing Subic Bay Naval Base and Clarke Air Force Base, and the U.S. presence was a sensitive political issue.

But the Phils had run into a problem when several of their Generals were caught in an attack on a Subic Bay headquarters by insurgents. The insurgents took the headquarters building and were holding the generals. The situation had to be resolved quickly, as a matter of national pride, if nothing else.

They took back the building successfully. All the insurgents were killed. The problem was that so were all the Generals. The Philippine Army figured they needed to do things a little differently, so they picked one of their senior Special Forces officers to set up their version of a Special Operations Command.

The General picked to run their SOCOM had worked with U.S. Special Forces and knew about C Company. He convinced his government to bring us down to train his hostage rescue force, albeit in secret. And so we got tapped for a classified mission into the Philippines.

Which delighted my guys. SF used to work in the Philippines before the politicians ran us out. We knew most of their SF troopers; everyone loved the country and people.

Many Americans retired and had families there, and many Filipinos have families in the U.S. Their SF soldiers were good and professional but normally worked in jungle operations, not Close Quarter Battle.

There was also a Communist insurgent group at that time called the New People's Army and other Muslim separatist groups in the south who wanted independence from the Christian majority population in the north. In addition to the political risk, there was some very real physical risk in being down there. COL Nick Rowe had been killed down there while I was in the Q Course.

But we were used to dangerous places, and like I said, we loved the Philippines and the Philippine people, so we went off.

Once we were settled in at Fort Magsaysay the General and his staff gave us our brief. In short, the request for our help resulted from the Navy resolving the attack on their headquarters alone. As I mentioned earlier, all the insurgents were killed, but so were the Generals. They figured they needed to change their approach.

I asked the General how they executed the attack. He grinned, "Well, they started by prepping the target with naval gunfire from the cruisers and destroyers in the bay. I explained to them later that this was probably not a good idea and that we knew a better way. And so, you are here."

OK, we can help with that. Naval gunfire isn't too discriminating.

We ran over training plans, equipment, and places to train. The training plans were easy; we had those on file and knew how to take people from start to finish. Equipment and a place to train would be a little harder.

At that time, the Philippine army had little body armor, no short rifles for close-quarters combat, or tactical lights for their weapons. Worse, they had no place set up for live fire close-quarter battle training.

For those who haven't done it, this takes a little explaining. Close Quarters Battle means just that. It's one thing to lie on the ground and shoot at people moving towards you from four hundred meters away. It's an entirely different matter when the bad guys are five meters away from you, there's more than one, and it's in a dark room. Things get interesting real fast.

The term Close Quarters Combat is also used, but that refers to individual soldiers fighting close up, hand to hand or with improvised weapons. The tactics and techniques apply to CQB, but CQB takes it to the next level. In CQB, you are attacking as an organized unit with firearms, explosives, and a tactical plan. It's more complex, and every member of the unit has to know what everyone else will do in any situation, and respond instinctively and instantly.

There are many tactics for entering a dark room with armed bad guys and good guys tied up and sitting among them. But it all boils down to doing something to distract the bad guys, even if only for a second, and attacking as a unit to quickly overwhelm all resistance. This means moving your team into the room before they can recover and shooting the bad guys quickly while not shooting the good guys.

Your actions are measured in fractions of a second and inches of accuracy. When shooting targets on a range, you aim for center mass. A hit anywhere on the target counts. And you get several seconds for each target.

When shooting CQB, you aim for center mass but need to hit center mass within inches. And you don't get an entire target to shoot at. A "good guy" target may cover the target except for a portion of the bad guy's head looking out at you.

In this case, center mass means hitting the spine or brain. The objective is to separate the brain from the hands so the bad guy can't shoot back at you or shoot the hostage.

Everything you hear about wound channels, knock-down power, etc., in the civilian ammunition market may mean something when you are hunting a deer, but it is meaningless when you are fighting an armed man shooting back at you. If you get a fatal shot on the deer but he doesn't drop right there, you can track him down and finish the job. Do that with an armed terrorist shooting back, and he will drop you.

So, you train with your rifle and your pistol to hit a three-inch target at ranges out to fifteen meters (about the size of a large room) in one and a half seconds. And you need to hit it twice in that second and half.

You need to hit it twice because, one, two bullets stand a better chance of killing the bad guy outright, and two, if he is wearing body armor with a hard plate, the body armor will stop the first round. But if you "keyhole" your shots and place that second round within an inch of the first one, the second round will blast right through the "keyhole" in the body armor where the first round weakened it.

Now, placing two bullets within an inch of one another in a target up to fifteen meters away in less than two seconds takes some practice. But it can be done. Doing that in the dark, using night observation devices or a flashlight attached to your weapon, is much harder, but it can be done.

Doing that in a small dark room where there are eight people all shooting at one another at the same time, and explosions are going off, and there are people in there you must not hurt, and you don't know exactly where they are, is damn near impossible. But it can be done. If you practice a lot and you have some talent for the job.

Training like this means you will be having bullets pass very close to you, even if no one is shooting back. The only way to clear a room of bad guys is to quickly cover the entire room and shoot the people who pose a threat. You have to get in among the people in the room, spread out, and cover everything with your weapons before the bad guys can react.

You designate sectors, and each man covers his sector, so life and death decisions are made within seconds, all at the same time.

Your buddy's sector overlaps your sector. You design your entry so that when you look left, your buddy looks at the area in front of you or to your right. If someone with a gun pops up right next to you while you are looking the other way, your buddy takes him down. This means that if someone pops up right behind you, your buddy shoots inches away from your body to hit him—very little room for error.

Fear and adrenaline are very real parts of this training. You just can't tell who is going to be too frightened to function or get too excited and miss something unless you put them into that small room with bullets flying past their ear and observe how they react.

A trained operator must be cool, calm, accurate, and fast. Every time. He has to know what his buddy will do before he does it. We called it "the flow." A well-trained team that has worked together will "flow like water" through a house in seconds, eliminating anyone with a weapon resisting their advance but not touching anyone or anything with their bullets that do not present a threat.

Actions are more instinct than thought. We call it "muscle memory." Your body reacts to situations before your brain has time to process what is happening.

One example of that is when your rifle runs out of ammunition. In a gunfight inside a house you want to stop after you clear each room, see how much you have left, and change magazines if you have to. But if you are in a gunfight with someone right in front of you or some hostages may be killed if you stop, there is no time to stop and assess or change magazines.

If you run out in the middle of a fight with an attacker in front of you, you drop your rifle, which then hangs from your neck by its sling, and draw and fire your pistol in one and a half seconds. And hit the three-inch center mass of the guy in front of you two times. In the dark. Again, it's not easy, and you usually move while all this is happening to make yourself a harder target to hit.

There is a lot more to this type of training, but it is and should remain secret to protect the guys who have to do it. But this should show you how it goes down and what you need to train to do.

Now, the Philippine Special Forces lacked the body armor and weapons needed to do all this, and that was a big problem. Not having a shoot house to train in was a bigger problem.

A shoot house is a building that has bulletproof walls that catch and hold a live bullet. This allows you to go in with small teams and shoot at targets like you would in real combat.

But building a structure like this isn't as easy as it sounds. Steel will stop a bullet if it's thick enough, but it can ricochet and could come back and hit you. Drywall and wood usually won't allow a ricochet, but it also won't stop a bullet unless there is a lot of it, and once it's shot, that hole won't stop the next bullet.

This requires some engineering and costs quite a bit of money, both of which the Phils didn't have much of at that time. So, we improvised.

This is where the "IRFs" are of real value to the CINC.

The IRF doesn't get the same amount of money or resources as national assets in the states. We have to do the best we can with what we have. The IRF is good at improvising, which the Phils needed.

I need to point out here that the training solutions that were designed were not my idea. Everything you read here came from the Special Forces soldiers on ODA 132 and their team sergeant, a great NCO who went on to become the Group Sergeant Major years later. They came up with the concepts and they did all the work. I was just along for the ride and to handle the coordination to get them what they needed to do the job. They were expert soldiers and trainers who could build training solutions out of nothing and accomplish what others would consider impossible without millions of dollars and years of time. That's what the "IRF" was really all about.

We set up a dry fire, blank fire shoot house in an old abandoned hospital. There were many halls, stairwells, doors, and windows where the tactics and maneuvers could be trained.

Marksmanship training wasn't a problem. Any regular shooting range will do. That's where you build muscle memory and get the guys shooting fast, accurately, and instinctively. Everyone stands online and practices the drills for speed and accuracy.

But they will never be ready unless you get them into a live fire shoot house, with bullets whizzing past their ears, explosions going off, and the certain knowledge that if they make a mistake, either they will get killed or they will kill one of their buddies. That's where you find out who can and cannot do this. And find those who can get used to that environment and work as a team.

Later, we would build a "tire house," the poor man's shoot house. You stack old tires one on the other and fill them with dirt. You create a row of stacked tires to make walls and then run another two rows of stacked tires positioned so the rows fit together and block the gaps between the stacks.

It works but takes time to build and must be done correctly. There are no rocks in the dirt, no empty spaces, and you need a lot of room. But you can shoot at targets 360 degrees around you in there. We didn't have time for this training iteration, the General needed some people trained fast for a demonstration for the general staff to prove his men could do the job.

After some debate, we set up three shoot houses with stuff we collected from a local junkyard.

First, we built a simulated house on the flat range using bamboo and walls made of woven palm fronds. This wouldn't even slow a bullet down, but as long as we placed the targets where the shooter would be facing down range when he saw them and went slow with one or two men at most, they could learn the basics of shoot-and-move training in a live-fire environment.

Next, we went to an abandoned ammunition storage area. There were large bunkers designed to contain explosions; concrete covered with dirt. There were two advantages there. First, if a guy makes a mistake, the bullet will not fly out of the range and kill someone. Second, we could close the door and make it dark.

We sandbagged the walls around targets set inside and built plywood partitions to simulate rooms. We were limited in what we could do, but we could run one, then two-man teams through, and ultimately four-man teams. Here, we could place targets on the sides and toward the back and get more movement and surprise built into the scenarios. It wouldn't last long, and we had to check sandbags after every run, but it would work.

The flat range drills, the bamboo shoot house, and the ammunition bunker weeded out those who were just not up to this type of training. We returned to the abandoned hospital with the ones we had left. This one was dangerous and very close to what a real target would look like. People lived nearby, and stray rounds could easily kill someone—those who trained there could not miss, just like in real life.

Here, we built wooden target stands that held a piece of steel from the junkyard that was about one by two feet and as thick as we could find. It was not AR500-hardened steel by any means, and it was not truly safe to shoot at, especially more than once, but it's what we had.

You want a very hard AR500 steel to ensure the bullet doesn't leave a dimple when it hits. If it does, the next bullet may ricochet in a direction you cannot predict. What we had wasn't that hard.

We added sandbags behind the steel and in front and covered the front of the box with a piece of rubber conveyor belt. This would stop bullets, more than once, and stop the ricochets from coming back at the shooter. At least for a while. We would build these for one training iteration when we practiced hitting a target in a real building but take them apart when we were done and throw away the steel.

This was the final exercise. The guys we trained had to move tactically through the jungle past populated areas without being detected, use explosives to blow the door into the hospital and clear the rooms, stairs, and halls. Targets were set up in various locations away from windows to ensure bullets didn't fly out a window and endanger the civilian population outside. The hospital was built of concrete and designed to stand up in a typhoon, so we were not worried about bullets going through the walls but if the troops missed the bullet trap, we would have ricochets whizzing around the room and it could kill someone in the room.

That's where you get the true stress situation, where the guy's training understand that if you miss, you could be killed. It's a good test for the instructors, too. You don't want to let someone do this training unless you are sure they are ready, since as the instructor you will be there with them.

Commanders get involved in this as well. The true test is the ability to kill the bad guys but not the hostage. One test is where the commander will sit at a table with bad guy targets behind him and have a team being tested enter the room and blast the bad guys behind you. If you don't trust your troops to do this without killing you, you have no business letting them do it with hostages.

The training went well, and about halfway through, the General invited me to lunch with him and his staff at his favorite restaurant downtown in Manila. Of course, I accepted. I love Philippine food, and the General is a great guy.

But during our time on Fort Magsaysay, we had gotten accustomed to a sense of safety and to some extent we had forgotten the risk of where we were. The luncheon brought that home to me. As I was getting into the van, a young lieutenant ran out and handed me a kit bag with something heavy in it.

He explained that it was our weapons but not to jostle it too much so the pins would stay in the hand grenades.

Really.

I looked at the General, who laughed and explained, "The last time we were there, we were attacked and it got a little too hot for comfort. I told my aide to throw some grenades in the bag this time in case we get caught in something we can't shoot our way out of. But with you along, we should be fine, right?"

OK, don't know if the General is pulling my leg or serious, but I'll go along. And I definitely won't jostle the bag.

Lunch went without incident, but we wouldn't leave the Philippines without finding out the General wasn't kidding. We had a final demonstration for the General Staff on the new capability the Philippine Special Forces could provide; all went well, everyone was happy, and the Philippine SOCOM became a reality.

To celebrate, the General organized a farewell dinner in Angeles City at another of his favorite restaurants. Again, this was fine with me. Angeles City, then and now, has a reputation for being a place where people go to have a good time, but it also was a place where the NPA had hit U.S. personnel more than once.

It sits right outside of what used to be Clarke Air Force Base when the U.S. was there and was a concentration of bars, nightclubs, and restaurants. The NPA knew it well, but then so did my guys. It was no more dangerous than many other places we went. We gave everyone the usual safety brief, know where you are going before you go there, stay in small groups, know where the exits are for an emergency, and don't sit with your back to a door, etc.

But when we went to the restaurant selected by our friends, several of the items on our usual safety checklist had been ignored. Granted, it was the General's show, and he wasn't obligated to follow our rules, but it did catch my attention.

A long table with nice china and candles was set up for us. Nothing wrong with that, but it was outdoors, in a courtyard right on the street. No wall, just a low row of shrubs about waist high. All the people passing by on the street could clearly see the formal table and all of us sitting at it.

To top it off, the head table with me and the Filipino officers was on the street side, and my back was to the street. The General looked at me with a big grin as I surveyed the setup.

"Sir, you sure this is a good idea?" I asked.

"My best men, the very ones you trained, have set up security. Don't you trust them?" He asked with a chuckle.

OK, General, you got me. Confidence test. No problem, let's do this.

We sat down and had the usual round of toasts and pleasantries until it came time for me to make a toast. I raised my glass to the General and rattled something off about eternal friendship and the brotherhood of Special Forces, glass raised when I saw the General's eyes widen at something behind me.

One of the things you learn working in bad places is to recognize subtle indicators that you are about to get your ass whacked. You sit with your back to a wall facing the door so you can check out people who enter. Body language tells you a lot, and most people coming to do you harm will give you a hint of what's coming. But you can't watch everywhere, so you sit with some of your buddies and everyone watches a different sector.

If your buddy's eyes widen in surprise, that's your cue to look that way, fast, and get ready to move.

That's the cue the General was transmitting loud and clear right now. Of course, this General also had a sense of humor and loved yanking my chain. He had directly challenged my confidence in the men we had trained. OK, I will play this through, and I raised my glass to his for the toast.

As shots rang out behind me.

The General was still looking behind me, but he wasn't ducking so neither was I. I just looked him in the eye and asked him, "Are we good?"

He shifted his gaze back to me, smiled broadly, and said, "We are now!"

We clinked glasses and finished our toast and turned to look out at the road.

In the road behind me lay a very dead man with a good-sized hole in him.

OK, this wasn't a joke.

But what the hell, I'm alive.

Many of our trips would go like this, working with many great people in bad places. On some trips, we would train on targets in major cities in the U.S. Always it was challenging and difficult. As with everything, there are more stories but sadly, most of them cannot be told.

And so we end my story about life inside the Army here. I retired from the Army when my second year in command of Charlie Company came to a close. I had twenty-four years in, I was only a Major, and the chance of me going much further was pretty slim.

Might have made Lieutenant Colonel if I hung around long enough, very unlikely I would ever get a battalion command, so Charlie Company was probably the most fun I would have in the Army. Everything from here on would just be hanging out to try and get another promotion. Not my style, so I decided to leave with my head up and try being a civilian.

Wouldn't work. But that's another story.

ODA 132 training in Hong Kong